CELLS AND SOCIETIES

CELLS

AND SOCIETIES

By John Tyler Bonner

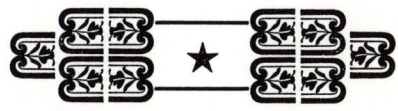

PRINCETON, NEW JERSEY

PRINCETON UNIVERSITY PRESS, 1955

CONTENTS

CELLS AND SOCIETIES

THE SAMENESS OF
LIVING THINGS

AFTER spending an afternoon at the zoo watching the baboons on a "monkey hill" it is always hard for me to leave when the time comes to go home. I know it is not because I am a biologist that my interest is so aroused but because I am a human being, and all the other fathers and the children and mothers watch with the same keenness. We watch as though we were looking through a one-way glass and we see them in their living rooms, their dining rooms, their bedrooms and their toilets; they seem to carry on a complete existence and are not abashed by our staring eyes. The old male jealously guards the females of his harem; the young coyly eager females in heat try to attract the attention of the males, the frustrated bachelors sit and watch, and perhaps the most consciously felt of all are the mothers tenderly caring for their small babies. They fight for food, they fight for females, they mate, they groom one another, they excrete, they play, and the mother helps the whimpering infant that cannot climb up the rock to join her.

I watch fascinated because I know without thinking that in all this buzzing activity I see my own emotions, my own desires and aversions. It is impossible to watch without projecting oneself into each situation and this is not just because baboons have human faces and postures; that is part of it, but it is mainly the actions themselves.

In this colony, even though it is rather artificially placed there by the zoo-keeper, there is a unity and an even more obvious complex interrelation among all the individuals.

For instance, there are the harem groups, each centered about a fierce old male, and the existence of all the other monkeys seems to radiate outward from these centers. But so as not to anticipate what is to come later I will not dwell now on their social structure, but say simply that by analyzing an animal society it may be possible at least to see its nature and its living properties more clearly, and then it can easily be compared to other manifestations of life. All living things have a certain sameness, and one does not have to scratch far below the surface to see this. We can learn some of the basic principles of life from animal societies, from a single cell, or from a multicellular organism.

To put the matter with utmost brevity, all animal societies have certain activities in common and of these we will choose three: a society takes in food so that its individual members may be nourished, a society perpetuates itself by reproduction, and a society has some coordination, some integration or communication between its members. The fact that all societies have these activities means that one can compare one society with another, and most especially, compare the methods in which they perform these activities. There is no reason to expect that they will perform them in the same ways, and perhaps the importance and the interest of the comparison is to see how the same activities can be performed in different ways.

Social groupings have occurred frequently in the animal kingdom, and we will pick examples from the higher animals. If we look next at lower forms we find that by custom we no longer dignify their groupings with the word "social" but instead use "colonial," and these colonies also take in food energy, reproduce, and are integrated. And soon it becomes evident that any living unit, whether it be a society, a colony, or an individual organism of one or many cells, they all, as a unit, perform these three activities. This is the road we shall now pursue; we shall compare the ways in which different units of life perform their living activities, and we shall see and perhaps to some extent understand the extraordinary variety of living things.

First we shall examine a series of mammal societies (starting in Panama with the howling monkey) which will lead directly to a consideration of some insect societies. Next, proceeding down the scale, will come the colonial polyps or hydroids and various types of single-cell colonies. The base line for this downward trend is to be the one-celled organisms that can perform the same living activities within one small membrane boundary, and then we shall turn up the scale again towards the multicellular plants and multicellular animals to see how within one organism these same activities are performed.

The initial section on animal societies will serve as a useful and familiar setting to introduce the fundamental problems, that is, the ways in which living things eat, reproduce, and are coordinated. In descending to the lower colonies and especially to the single cell it is possible, because of the simplicity of the organisms, to examine the fundamental meaning of these three living activities. Then, armed with an understanding of their existence and some insight into their nature, the reader will be prepared to see how in the internally complex multicellular plants and animals these living functions are carried out.

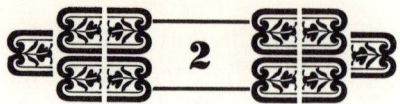

HOWLING MONKEYS

The air has a quiet chill (with a faintly antiseptic smell of mildew) and across to the hills the grey mist lies low and heavy over the grey-green jungle in the light of dawn. Gatun Lake, the central part of the Panama Canal, stretches out below, the water grey and slightly rippled, and behind the buildings is the deep forest of Barro Colorado Island, noiseless save for the dripping of the dew and mist from the trees.

And then the sodden stillness of the air is broken by a long, low guttural roaring wail way off in the jungle. This is followed by another and then another in a different direction and this means that soon the mists and the quiet will dissolve and the hot sun will rise, for the howling monkeys are giving their calls and have started their day.

In the jungle the tall trees vault together in an airy ceiling like a giant aviary cage and the sun filters down in small shafts to the ground—too little for grass and shrubs to do much growing, and the ground is mainly bare but for the dead leaves and humus.

Along the thin trail all sorts of animals begin to show themselves or make their noises to break the heavy quiet of the forest: sometimes a bright butterfly, sometimes a colorful toucan on a high branch, sometimes the grunts and scamperings of a herd of wild peccaries. The jungle is alive, and with patience one may see coatis, armadillos, squirrels, marmosets, and many others. They show themselves and chatter or timidly disappear.

But should one wander near a band of howling monkeys this peaceful scene suddenly changes, for out of nowhere an overpowering explosive roar numbs the air. This is rapidly followed by another and soon all of the males in the band are angrily roaring. And then the smell in the air becomes rather acrid and unpleasant from the streams of urine and faeces raining down on the ground all about. There may be a dozen or more monkeys: the large terrier-sized brownish-black males, the slightly smaller females (Plate I), some of them mothers with young clinging to them. If a group has been startled as this one has, they will not continue their normal ways until they are again left alone, and then the roars slowly die down and they resume their social activity.

Perhaps we might still know very little of this social activity were it not for C. R. Carpenter, the authority on monkey societies, who for a period of two years spent much of his time virtually living with the howling monkeys watching and recording and attempting to understand their ways. Sometimes he would try to find a natural hiding place, or sometimes build a blind so as not to disturb the monkeys. He found that after he followed a group for some months, and they obtained frequent glimpses of him, they became used to him and would no longer roar and react so violently to him. On Barro Colorado Island there are some laboratory buildings as well as sleeping quarters, and one special band of howlers, called the laboratory clan, would repeatedly come very close to the buildings. This clan in particular showed the least signs of disturbance in the presence of human beings.

For a band of monkeys the principal occupation during the course of a day is eating. I say principal for this desire seems to govern their periods of rest and activity, and their movements as a group. They feed on the fruits and leaves of many different trees, and since they are satisfied by a wide diversity of plants it is hard to imagine their ever having a shortage of food. The abundance, in a tropical rain forest, of fresh green succulent buds and fruits hardly ever fails, except in a severe drought, and they are rarely subjected to any real competition or struggle for food. This fact is emphasized by their wasteful eating; they drop about a third of what they pick.

If they have chosen a good eating tree for their night's rest they may begin sporadic eating shortly after dawn, but more frequently they will first move to a desirable tree and start a long and leisurely feast of two hours or so. Towards the close the monkeys will one by one become full until they finally all rest or sleep, from the mid-day period until the early afternoon. The young may not be so eager to siesta and often frolic about while the elders remain quiet. Then the afternoon feast starts and may continue until six o'clock in the evening, when they then proceed to a suitable "lodge" or sleeping tree to spend the night.

The whole process of feeding, although unhurried and deliberate, is entirely dependent on the locomotion of the monkeys. An individual howler monkey walks along a branch on all fours, clasping the branch with his hands so that the thumb and the fore-finger are parted from the remaining three fingers, but his feet have the thumb opposing the remaining four fingers, much like our hands. His tail is prehensile and curved, ever ready for grasping, and is actively used in more difficult climbing; he also frequently uses it for hanging while eating, as well as for a fly switch and a grooming tool.

The movement of an individual monkey is usually rather sedate, especially when compared to the more active spider or capuchin monkeys. The howlers do not leap madly through the air or take great plunges; they progress with caution. This is especially evident when the howlers pass from one tree to another, for they will find the two closest branches for their bridge. Should the gap be wide, an adult may swing one of the branches to bring them periodically close together, and leap at just the right moment. It is even more interesting to see a full-grown monkey help a young one across a gap. The adult will grasp one tree with his hands and the other with his feet, thereby forming a bridge so that the young one can easily cross over his back.

Howling monkeys live almost their entire lives in the trees, and in all the time Carpenter spent with howlers he only once saw them jump up from the ground near a stream, and there are a few other scattered accounts. It is also known that they can swim if necessary, but such observations are equally rare.

The fact that howlers always move from one tree to another by the safest route is very helpful to the observer, for it is possible, by hiding near such a travel bridge, to see the animals come over one by one, making it a relatively easy matter to make an accurate count of any particular group. One question studied carefully by Carpenter was whether there was a definite sequence in the chain of individuals that passed from one tree to another. By watching numerous groups, and each group many times, he found that there was tremendous variation but that on the average the males tended to be first (although no particular male sequence was constant) and the females with young took up the rear of the procession. It was not unusual, though, to find a female, even one carrying young, leading the troop.

Barro Colorado is an island about three miles in diameter, although perhaps because of its irregular shore line and hilly terrain it is more accurate to say that it contains a little less than four thousand acres. In this small area Carpenter was able to study not only one or a few groups, but all the howler clans on the island. He found that the clans varied in size from 4 to 35 monkeys, 17 being the average. Each of these clans had a territory which it would defend and keep free of neighboring clans. The first reflection of this grouping was shown in the early morning howling, for by their howling each group probably gauges its position relative to neighboring groups. The monkey "countries" or territories themselves have only approximate boundaries and they will overlap provided the two groups are far enough apart. If, as occasionally happens, the clans come close together, there is a tremendous prolonged roaring on both sides in which all the males of both clans appear very excited and very angry. Even the females will bark and the young whine. But the opposing clans never actually come to grips. It is a vocal battle, and after a while the groups will separate again and become quiet. Animals frequently rely on bluffing for such warlike activities as defending territories, and this has the obvious advantage of avoiding bloodshed and loss of life. The great roaring of the howling monkeys, however noisy and unpleasant, is in the final analysis a harmless affair and yet at the same time it does effectively serve to keep the clans apart.

In defending a territory the males not only preserve their food but also their females from marauders. Reproduction (in the sense of the production of new individual monkeys rather than the formation of new clans, which will be discussed later) is an important binding force in the structure of the clan.

The females are not receptive at all times but only during short recurrent periods of heat which may last two or three days. When a female in the clan begins such a period, one of the males may show great interest in her. The two monkeys face each other and rapidly and rythmically push their tongues in and out and up and down. The female will then crouch on all fours with her rump high and accessible, and the male will quickly mount and give a series of rapid thrusts, the whole process of copulation lasting only thirty seconds. In ten minutes or so they will again mate, although the interval between copulations soon lengthens, and as this happens the male becomes progressively less aggressive and interested, while the interest of the female is unflagging. Soon she is seen starting the tongue movements and placing herself before him, and finally he shows signs of being utterly satiated and will either walk away or snarl at her advances. Now she will choose another male among the clan and he will remain her consort for as long as he is able. There seems to be no original proprietary feeling connecting any particular female with any particular male, but the interchange is quite complete among the clan. Once a female mates with a male, however, she continues with him until he has done. Carpenter once observed two males approach a female simultaneously, and the male that was the fastest and copulated with her first remained her mate for many hours. It is also interesting to note that frequently when there is a great commotion and roaring between two clans that are having a territory argument, in the excitement some of the monkeys are sometimes seen copulating.

Unlike many social animals, the male is not a dominating autocrat, for the clan is based on a system of equal sharing and exchange. Occasionally some of the older males will show a slight dominance, but nothing compared to some other monkeys. In Zuckerman's account of the Hamadryas baboon, mentioned in the first chapter, the males have harems, and depending on their supremacy in a colony they may have one or many wives. There are also many bachelor males who may be associated with a harem, but they are not allowed to touch the females and the fierce overlord sees that they do not. In captivity the struggle for power, that is, wives, is so violent that many deaths occur, and most frequently the victim is the poor subdued female over whom they are fighting. Furthermore, if food is thrown into a Hamadryas baboon family, only the master eats while his wives and children and the associated bachelors dare not even make a move toward the food. The others are allowed to eat only if there is abundant food, and the master has finished. By comparison the howling monkey is both modest and gentle; there is more than enough food for all and no struggle for it; there are enough females to satisfy all the males sexually, and even if there were not, their generosity in sharing might well compensate for the lack.

When an infant is born it is assiduously cared for by the mother for long periods of time. Carpenter describes an actual birth where a clan was in the process of moving from one tree to another, and suddenly in the rear of the troop there was a peculiar noise that ended in a half-grunt half-groan. The forward males immediately stopped and turned back and the mother was then seen with a small and very wet greyish-yellow infant. For about an hour, until darkness came, she could be seen licking and grooming her newborn, occasionally holding him up by the tail to clean his hind regions. Other females surrounded the mother and attempted to examine and touch the baby, but the mother turned her back and quietly tried to discourage their interference. An infant will immediately cling to his mother, first on the mother's belly, holding onto the long hairs of her sides. But after a month he becomes slightly more agile and rides on the lower part of her back, which greatly facilitates her free movement. He begins to make short excursions away from her, but at first they are very short and she is always watching carefully, sometimes even holding him with one hand. Only after six months will he wander by himself, and even then never far from his mother. Weaning does not take place until eighteen months after birth, and the young howler cannot be considered completely independent until he is about three years old.

The matter of how new clans are formed is unfortunately more speculative. The groups vary considerably in size—from 4 to 35 individuals—and presumably when a clan reaches a certain size it may split in two. Carpenter has often seen a large clan become temporarily divided, and perhaps such repeated temporary fissions finally result in a permanent split. In two successive years (1932, 1933) Carpenter made a census of the island and found that there was an increase from 23 to 28 groups in one year, which presumably must have occurred by such splitting. It is interesting, although puzzling, that recently two zoologists, Collias and Southwick, have made another census of Barro Colorado and find 29 groups, but instead of 17 as the mean number of individuals in a clan, as Carpenter found, it was 8. It is known that howling monkey populations may be decimated by jungle yellow fever, and they suggest that this great decrease in population may have been caused by recent disease, but the evidence is rather uncertain.

One curious fact revealed by these population counts is that the number of adult females was more than double that of adult males. The cause of this is not known: there could conceivably be a higher incidence at birth of females, caused by such factors as higher uterine mortality of males, or it may be that for some reason infant mortality in males is greater. Another factor that could account for some but not all the disparity is the fact that frequently males become isolated from groups and wander about as single bachelors in the forest. Sometimes these bachelors try to enter a clan and they are vigorously rejected and roared at by the males of the clan. But it is assumed that if they keep following a clan for long periods of time at a safe distance, that distance may become ever closer until finally they are accepted by the group. The wanderings and regroupings of such solitary males would certainly serve as an excellent mechanism for avoiding excess and continuous inbreeding in a clan; they would serve as a sort of genetic bridge between the clans.

Of particular interest are the various means whereby the animals communicate with one another, that is, the means of coordination, of integration of the clan. The most obvious of these is the howling and the various other types of grunts and barks they make: can these actually serve as some sort of primitive language? If one would consider as a language talk in which merely a dozen different bits of information could be transmitted, then this is a language, but it is greatly limited if one compares it to the vast and subtle languages of man.

Since, according to Carpenter, there are at the moment only nine known symbols in howling monkey language, it might be easy and profitable to quickly learn their language and especially to see in what ways it does integrate the group. In the first place any disturbance of any sort: man, a bachelor male, an aeroplane, or even rain or the wind brings out the low resounding echoing howl in all the adult males which we have already talked of frequently. When this happens the females give a terrier-like bark that can only be heard short distances and this concert immediately throws all the members of the clan into a state of anxiety and anticipation, ready for defense or offense. It also serves, as was said before, to mark out the territory of each clan, and even serves as a substitute for actual bloody fighting. Another sound, found in the adult males, may almost be considered as a prelude to the full roars; it involves "gurgling grunts or crackling sounds" and indicates mild apprehension or alertness to danger. The other males of the group immediately pick up the grunts and become, as it were, coordinated for action. There is even a third type of sound that goes with strange and new situations, a grunting that sounds like "who, who, who." But these grunts are milder and all the clan seem to show interested attention rather than fear or aggressiveness. It is important to realize that all these defense sounds are very helpful to the group. Each individual, especially the wary females with their young, is so to speak a sentinel, thereby increasing the eyes and the ears of the group many fold and thereby insuring greater safety. It is a cooperative effort which obviously has protective value.

A similar integration is seen in a fourth type of speech which serves to direct the group along the most favorable path, and to keep it together during its wanderings. When the possible tree routes are either numerous or poorly defined, all the males wander about apparently exploring for the best trail. If any particular male should find a convenient pathway he gives a repeated metallic cluck, and almost immediately the whole clan begins moving in his direction. This same sound is used to start the progression if the animals have been feeding in a tree, and continues to be used to direct and coordinate the troop movements.

Two more sounds are specifically associated with a young monkey that falls to the ground, an occurrence which has been observed a number of times. The young one gives a series of high-pitched tones usually of three successive notes and the whole clan swoops low in the branches searching for the lost infant. The mother has a special cry, "a wail ending in a grunt or groan" as she searches desperately. Howling monkeys, as was said before, have a great aversion to descending to the ground, but finally the mother leaps down, quickly places her child on her back, and climbs rapidly to the tree-tops with the rest of the group. Once Carpenter shot a female and her infant fell unharmed to the ground. One of the males swooped down, put the child on his back, and climbed aloft.

The remaining three sounds (to complete our nine) are associated with childhood. A very young monkey occasionally purrs like a cat for several seconds and his mother responds immediately by cuddling the child. Young animals playing actively often give a small "chirping squeal" which seems to be an expression of youthful exuberance as they bounce about. Frequently they hang by their tails, two or three at a time, wrestling with all fours and playfully biting. Sometimes this play becomes a little too boisterous, and with rising tempers the little ones begin quite a commotion. But this is suddenly and effectively put to a stop by the grunting of an old male. This sharp signal of authority either stops all activity of the chastened youngsters, or they continue to play in a more gentlemanly fashion.

The voice is not the only means of communication

among howling monkeys; equally important are movements and attitudes. For example, as we have already seen, tongue movements and the assumption of inviting postures communicate the sex desire. Another example is seen in aggressive rage, for often not only will an adult male roar, but will violently shake branches, break off dead limbs, and sometimes rush about; all to supplement the ferocity of the roar.

And then little domestic scenes, not unfamiliar to us humans, may often be seen. For instance, sometimes a female will have a second child before her older one is fully grown and independent. He has the eternal problem of the older child and resents this small thing clasped to his mother's belly. He may be seen trying to pull the baby away, and at other times, perhaps as the day is drawing to a close and the howling monkeys are curling up to sleep in a lodge tree, he may be seen trying to wedge himself between the baby and his mother, to regain that coveted warm spot in which to cuddle and sleep the long cool night, when the mists will again rise over the black forest.

SOCIETIES AND EVOLUTION

In a moment's reflection we can see that not only does a clan of howling monkeys feed, reproduce, and show a certain coordination, but that it could not exist at all if it did not do these things; the very life of the clan depends on these activities.

The necessity of eating is especially obvious; the howling monkey converts the energy of the carbohydrates, fats, and proteins of the leaves and fruits into its many activities. Much of the energy is turned into muscle movements and a small portion is used in the nerves which coordinate these movements. But feeding involves more than just the final conversion of energy into body activity for there is a whole complex of interrelated events. There is chewing and digestion as well as climbing into the trees and grabbing fruit. And to complete the circle these activities in turn are dependent on the energy from the food.

As monkeys die from old age, accident, or disease, they must be replaced by reproduction or the colony would soon dwindle to nothing. The fact that the reproduction of individuals is sexual, involving the fusion of the hereditary make-up of a male and a female and not by some sort of asexual budding, can be understood in terms of the mechanism of evolution. Each time an offspring is produced from two parents the offspring will have a new set of hereditary characters, a recombination of genes not found in either parent. And since the progress of evolution is dependent on the natural selection of variants, the sexual process greatly increases the number of variants and keeps the hereditary patterns in a fluid state. Sexuality serves as a great aid in evolution, or perhaps the matter may be stated more simply by saying that if a species must become adapted to a new environmental situation, it can do so far more effectively and rapidly by sexual reproduction than it could by any asexual process.

Sexuality itself has certain very definite effects on our monkey society. In the first place there is the physical segregation of the two sexes, and associated with this sex difference there are certain divisions of labor: the males assume a protective role by howling and tree-shaking as well as path-finding and troop-leading, while the females play a more passive role and center most of their activity on raising their young. The process of sexual union, the rearing of the young, and the defense of the clan are all patterns of behavior which serve to integrate the society and are all related to sexuality.

The relation of the rearing of children to the social existence of animals has been much discussed in the past. Frequently the young of a social animal are relatively helpless and need considerable care for long periods of time; but this does not mean, as has been sometimes claimed, that the cause of the social grouping is the protection of the young. It is conceivable that in some cases this might be a factor, but it is equally likely that with the integration of a society there was no urgent need for independence of the young, and therefore changes in evolution which led towards helplessness in children were not eliminated by natural selection. Even among relatively non-social, solitary animals there may be a wide divergence of helplessness in young; the naked and blind chicks of a robin require long periods of tender care before they can face the world, while a duckling will within a day after emerging from the egg reel out of the nest, plunge into the water, and swim about like a practiced master.

The need for coordination in a society is not quite the same as the need for reproduction or feeding. If by some freak of nature, say a windstorm, or by a man's collecting animals in a cage, a number of relatively solitary animals are artificially grouped, they do not by this device alone become a society. A society exists only when the group is to some extent integrated or coordinated.

Coordination manifests itself in many ways. In the howling monkeys we saw that by a combination of calls and actions the individuals within a clan communicate with one another and respond to one another. And furthermore this communication not only extends within the group, but among the various clans as well. In these monkeys this coordination is apparently entirely mediated through the nervous system of the individuals, which are connected to one another by a vast network of signs and countersigns.

Feeding, sex, and communication are hardly the only activities of a society. There are many minor ones which at first do not seem to be indispensable, but when considered in the light of evolution their importance grows. For example, consider the activity seen especially in mother howling monkeys, of helping their young across difficult passages from one tree to another by holding the branches and allowing the young to cross their bodies. Were this instinct suddenly eliminated from a clan, by some odd mutation, the clan might get along somehow. The infant mortality would probably rise but let us assume that its rise would not be enough to wipe out the clan immediately. But now this clan, and the daughter clans that spring from it, will over a long period of time be in competition with other clans that still retain this infant-helping instinct. In the long run, and all other things being equal, the clans with this child-helping instinct will be more easily able to keep up their population. That is, over a great period of years, there will be more offspring descended from those that possess the instinct than those that do not, because in the latter group the infant mortality is slightly higher. Natural selection, then, tends to favor the reproduction of monkeys having this instinct over those that do not, and so this child-helping is essential in terms of survival by evolution, but not for the immediate existence of a clan. Many favorable actions are encouraged, and should new traits appear in the monkeys, they will, depending on their nature, be selected for or selected against. This is the essence of Darwin's theory of natural selection.

As we see from this example, natural selection acts both on individual monkeys as well as on the whole clan, and now we may ask how is it possible that animals such as the howlers find a social existence advantageous and successful in competing for a place in nature? There is no doubt that animal societies have been a success; they have arisen independently many times in the course of evolution, and in some cases are known to have remained stable and abundant for millions of years. The question then is what advantages emerge for animals grouped together in a society?

In the course of many years W. C. Allee, a zoologist at the University of Chicago, has discovered and brought together instances where the mere grouping of animals may be mutually beneficial. In some cases, because of some sort of chemical produced, certain single-celled animals (and other forms including fish) grow faster in more crowded conditions. In another case he showed that flatworms grouped together were better protected from the killing effect of ultra-violet light than were isolated worms, and he was able to show that this protection was more than could be expected from the worms shading one another. But it is seldom true that increasingly large groups become more and more beneficial, for there is usually a critical or optimal size, and in some instances this may be a fairly small group.

A good example of the advantages of grouping is seen, as Allee points out, in the work of the British naturalist F. F. Darling, who shows that in the nesting of the herring gull in gull colonies off the coast of Scotland, the greater the size of the colony, the greater its chance of rearing a large per cent of young gulls relative to the total population. The hypothetical argument is rather simple and ingenious. If there are many gulls present there is by mere numbers alone a mutual stimulation to begin mating early, but even more important, a stimulation for all the birds to begin at approximately the same time. This means that in a large colony all the young gulls hatch at the same time. The newly hatched birds covered with down are very vulnerable, especially to foxes and other predators, and by bunching this period in time they get through it all together, before the predators can do much damage. The small colonies, on the other hand, continue to produce young throughout the season and provide a constant source of food for the predators.

Another way in which grouping clearly helps is by helping to bring the sexes together for reproduction. This is especially obvious for those animals which come together only during short periods of the year, during a mating season.

Still another example is provided by the howling monkeys in the fact that the group has many eyes for sentinal work. Should any one of the monkeys see a presumed foe, he will give the signal, and all the others will instantly become ready for action. Also in path finding (but this is perhaps a minor point) each male looks for a good trail, and the successful one calls for the rest to follow him.

In all these cases, and there are many others, there appears to be some advantage in numbers, but even more important is the fact that frequently the *integration* of the group gives the advantage and not just the grouping itself. For the howling monkeys the important fact is that information is immediately passed on to the whole group, and they all respond instantly.

Because integration is successful and useful to a society, during the course of evolution those variations or adaptations which tend to give the group a unity will have an advantage, and the offspring of those groups may tend to become more numerous than less well-organized groups. The integration which characterizes societies is something which natural selection will tend to keep and nourish, but there is no law which says how this integration takes place. In the howlers it was largely by integrated behavior patterns, and this is true of most higher animal societies, but still there exist great variations, as we shall see.

In the lower forms, as one descends to primitive groupings of cells, a new and surprising fact appears. The integration is no longer dependent on instinct and behavior but on other interesting mechanisms, and it becomes obvious that these colonies of cells are not producing a cell society, but ultimately an individual multicellular organism.

Thus we find that not only do animal societies eat and obtain energy, reproduce, and show some degree of coordination, but so do lower colonial organisms, single-cell organisms, as well as individual multicellular animals and plants. Living units, whether societies or individuals, have the same basic functions for existence, and they are affected by the same laws of natural selection. Natural selection works in terms of units, and these units may be individuals, colonies, or societies. Selection acts in this way because units perform certain functions that foster survival, and we may therefore call a society or a single organism a functional whole.

But these functional wholes may achieve their wholeness mechanically in a wide variety of ways and we are now interested in that variety. Howling monkeys get their energy differently from army ants, which in turn are different from protozoans or green plants; but all need energy, all must reproduce, and all must be coordinated.

FUR SEALS

THE cold, barren and raw Pribilof Islands lie north of the Aleutians between Alaska and Siberia in the Bering Sea. These islands are five in all, the two largest being Saint Paul and Saint George, each about seven miles long. They were discovered—desolate and uninhabited—in 1786 by Gerassim Pribilof, a navigator for a Russian trading company. His company, like many others, was seeking furs, and here on these small islands he found fabulous numbers of fur seals. But soon the Russian nobility and the imperial family learned the value of these seals and assumed the sole seal-killing rights on the islands. When Alaska was sold to the United States in 1867, the Pribilofs were thrown into the bargain and immediately various American companies began vigorously exploiting the seals. In spite of poor methods, the Russians after a period of time managed to keep up the size of the herd, even though in a year they would take as many as 70,000 seals. But in their first year the Americans killed approximately 239,000 seals. This unrestricted killing, and the wasteful killing of seals at sea by small schooners equipped for that purpose (this was the profession of Jack London's "Sea Wolf") soon put the very existence of the seals in danger. Then fortunately the United States Government in 1896-1897 made a study of the situation, including a careful analysis of the habits and ways of the seals, and by a combination of treaties with other countries to stop seal-killing in the open waters, and with the government itself doing the killing on the islands, the population has steadily increased and is at present flourishing. According to a recent estimate the herd consists of about one and a half million seals (approximately the population of Washington, D.C.) of which about 65,000 are killed annually, yet the colony, under careful supervision, is maintaining its population level.

During the winter the fur seals migrate great distances, but each summer they return to the Pribilofs. There they find rocky and sandy beaches sloping gently into the sea making it easy for them to do their clumsy waddling out of the water. The summer temperature is between 40 and 50°F, with mist and fog and rain almost constant. It is a rare and momentary sight to see the sun, and when it does come, all the seals lie as if prostrate with the heat, and thousands and thousands of flippers will be seen waving in the air, as though cooling themselves by fanning. There is not a tree on the islands, even near the neat white salt-box houses of the Aleut villages, although the ground inland is covered with a mat of green grass. To man it seems a lonely and forsaken place, exposed and harsh and nude like its cold black rocks, yet to the seals it is a friendly gathering place where they may feel the warmth of company which they seem to crave. All along the beach and even far back in shore there will be groups huddled, sometimes the harems of old bulls, sometimes a "pod" of young pups, to use the sealer's terminology, sometimes a group of frolicking young bachelors. The noise is deafening, for with the sound of the surf they are roaring and barking and wailing constantly all through the day and night. And to a human being the stench is powerful for in their ungainly inch-worm waddle they plow through the litter of the beach, the excrement and sometimes the dead pups, taking no heed of their surroundings, but showing interest only in the other seals.

One of the most striking aspects of the appearance of fur seals is the great disparity in size between the sexes. The female or cow is about 4 feet long and weighs about 70 pounds (Plate II). The male on the other hand is a good 6 feet in length and weighs about 600 pounds, and there are even instances of bulls approaching 700 pounds

(Plate III). A cow will reach her full size in three years, while a bull takes seven years. For the first three years he closely resembles a female; then in his fourth and last year as a "bachelor" he begins to grow a "wig," the long bristles down the back of his neck, which incidentally make him and his elders valueless as pelts. In the fifth and sixth years he is a "half bull," until finally he becomes mature in his seventh year. He may then gather some cows and become a "harem bull," or if he is not bold enough he may become an "idle bull."

To come now to the matter of their feeding, of their intake of energy, fur seals are fish eaters and do all their eating in the sea. They eat pollack, salmon, cod, lamprey, squid and many others. The seals are agile swimmers, and by diving and plunging and putting on great bursts of speed all these smaller delicacies are gracefully caught. Even though there are a great many seals, the ocean is virtually an endless cornucopia of fish and, as with the howling monkey, there can be no question of want of food. There is no competition; it is merely a matter of making a little effort to find and pursue the food.

When the seals have not hauled out for the summer on the Pribilof Islands they are constantly in the open sea, wandering sometimes in small groups of five or six but often separate and isolated. They explore and they play, but mainly they feed and follow the schools of fish, becoming fat and laying down a thick layer of insulating blubber. Exactly why they migrate great distances is not clear, but possibly it is because of the fish they seek, or more likely, as the winter comes on they seek warmer water. When they leave the islands in the late fall they swim rapidly southward along the North American coast and are found anywhere from Vancouver almost to Los Angeles. In April and May they move up again to the Alaskan shores; in June they are along the Aleutians, and then finally they come to the Pribilof Islands during the summer. This information was obtained by the oceanographer C. H. Townsend who examined the logs of the so-called pelagic sealing fleet and found the kill, the date, and the position from the records of 123 vessels. These boats, when sealing in the open sea was permitted, would kill as many as three hundred in a good day. The most favorable way of catching them was when they slept lying curled up at the surface of the water with their snout and their hind flippers sticking out, the flippers curved back toward the snout. Often they would be playing or rolling about in large masses of floating seaweed and showed little alarm as the large canoe of the hunter with his harpoon or shotgun approached.

When the drift ice which packs about the islands goes in late April or May, the old males come in first and haul out on some favorable spot on the rookeries. They remain in their territory throughout the breeding season; sometimes they stay there for two months and during all that time they do not eat and hardly sleep, yet they are undergoing the violent exertion of maintaining a harem. The old bulls come in sleek and fat in the beginning, and at the end their blubber is gone and their skin hangs loose and scraggy for they have lost about two hundred pounds. They have managed to eat enough during the long winter to last them through a mating season—they take their pleasures one at a time. The cows may go without food for a week or two, but never more than this for they slip into the sea periodically and return to give milk to their pups.

The whole society of fur seals is primarily centered about reproduction. The females must give birth to their pups and rear them on land, and at the same time, in this great and convenient aggregate of seals, they are fertilized so that each may bring forth a pup the next season, for the gestation period is about a year. The work at the Pribilofs is one of making and rearing new seals; it is an intense and violent activity crowded into the few warm summer months.

The old males which come first often occupy the same

territory on the beach that they had the year before. They may fight for their territory and no doubt there are frequent changes of position. The idle and half bulls may be about in the water but they do not haul out until the arrival of the cows in June. In the beginning they stay close to the water's edge and try to intercept the cows as they come, and later they may be found in fair numbers in the rear of the rookeries, always looking for a chance to abduct a cow.

The younger bachelor seals arrive early also but they keep quite separate from the harem bulls in the "breeding grounds," and go to their own private quarters, the "hauling grounds." They may be on sandy beaches close to the rookery, or they may be in flat areas above and behind the breeding grounds. Sometimes, to allow the bachelors to get to their rear positions, there is a neutral strip or runway so they can come and go unmolested. The seals that are killed are nowadays driven inland from their hauling grounds and in this way it is easily possible to separate the young males which are used for furs. Later in July the two-year-old seals appear and soon afterwards come the yearlings.

The cows come in very gradually in June, and the first ones will attempt to come unnoticed alongside a bull. If he sees her he may try to catch her, but she shies away and slips back into the sea. He may actually grab her in his mouth and dump her by his side, but if she escapes, she will come to him of her own accord later. The next cows to arrive apparently have a strong desire to join the first cow; the urge for company is most compelling. This means that the harem of one fortunate bull will rapidly swell while those in the vicinity will be alone. Sometimes the size of these initial harems will go over one hundred cows, and the bull will, by constantly herding them together, be able to keep them in one tight group. Some hours after the cows haul in, they have their pups and five or six days after that they come into heat. When this happens all the bulls become very excited and the din from the rookery becomes increasingly greater. The original bull can no longer keep the other bulls completely away, and usually the size of the harem drops so that it may be finally of fifty to one hundred cows. But the cows are not the same during the season; new ones keep coming in and the ones that have been served spend much of their time at sea.

Fighting may take place between harem bulls or between a harem bull and an idle bull that is attempting to literally carry off one of his wives. If such an attempt is made the old bull will bellow and roar, and instead of daintily circling around his harem he will plow directly across them, sometimes walking over the new born pups, and even occasionally crushing them to death with his great weight. The two bulls, now face to face, will bellow and make mock lunges at one another, and if neither is intimidated by these antics they begin to tear into one another by repeated lunges, ripping open the flesh in great red wounds about the neck. If, for instance, the marauding idle bull is turned over he may then try to escape, but now, before he gets back to the safe rear grounds, he must get a further severe punishment from the other neighboring harem bulls, who are always happy to pounce on a man when he is down. Sometimes the idle bull will be caught with a cow in his mouth and then they may end by tugging on her and ripping her apart. Sometimes the idle bull will escape with a cow, but the harem bull will not chase him far, for if he does, bulls from the other side will be quick to dip into his supply of cows. The whole course of this fighting is quite a different thing from the harmless bluffing of the howling monkeys. There is a strict order of dominance here among the bulls, and like the Hamadryas baboons of Zuckerman, the fighting is in earnest and the wounds serious. It would seem that this destruction is unnecessary and wasteful, yet obviously it has been well compensated for in nature by a sufficiently high birth rate. It is possible also that most of the wounds seem more serious to us than they do to the seals. They are protected with a thick layer of blubber and they do not seem to notice even a severe wound; they never cry in pain and they never lick the gashes. Blood seems to evoke no reaction in them for they walk through it, swim in crimson sea water, or become covered with it by a deep wound and in no case show any interest.

When a cow comes in to have her pup, she first shows the characteristic difficulty of advanced pregnancy. No matter how she lies, every position is uncomfortable, and this restlessness continues until the pup is born. The seal has but a single pup, and since this pup occupies only one horn of the uterus, the other horn can be almost immediately ready for the reception of the next embryo. In this way fertilization can take place rapidly after childbirth and then the new embryo lies dormant for a considerable time, apparently a mechanism to prolong the gestation period. Both of these mechanisms are obviously very advantageous in such a short social season.

The pup may be born among boulders, on the sand, or in a crevice between two rocks. Seals, with their soft blubber, do not seem to be bothered by lying on the rocks, and this is true during the *accouchement* as well. When the pup is born the cow will quickly turn about and break it free from its caul and bite off the cord. Often the pup will keep sliding away from her on the slippery rocks and she will pull him back, or she may take him in her mouth, as a dog will, and find a better place for them both. Then she repeatedly pushes the pup to her teat to try and get it to suckle, but it takes some time and some perseverance before he finally will eat. Sometimes the cow will not show the proper attention and leave the afterbirth attached. This will eventually drop off but cases have been observed where the afterbirth is caught in a crevice of some rocks and the luckless pup is fettered by his own umbilical cord.

The pups stay on land for a few weeks, getting fat on mother's milk, and then they begin slowly, by degrees, to take to the sea. During the early part of their existence they are subject to many dangers, and are often found dead or mutilated. Frequently they die of worm infestations, or they are starved to death, apparently separated from their mothers. The mothers return to the sea and come back every five to ten days to feed the pups; sometimes the mothers do not return or they cannot find the pups in the great crowd. Crushing, especially by bulls is a very frequent cause of death. If a bull walks directly over a young pup it will most certainly die, but their resistance to crushing is remarkable, for often it seems impossible that they could survive after a bull has rushed past, but they scramble away apparently unhurt. An instance was seen where a bull was copulating with a cow and all the while there was a struggling pup underneath. He bided his time until they were through and then scrambled away quite safe, at least with no external signs of trauma.

It is very difficult to find an account of the actual courtship of the seals, for the government publications, while very detailed concerning all their other habits and behavior patterns, are extraordinarily circumspect and Victorian on the subject of sex. The most that is revealed on these official pages is "copulation was observed at 8:47 and 9:31 P.M." and then the report hurries on to another subject. Instead we must go further back to the early account of George William Steller, published in 1751. Steller was the famous naturalist who accompanied Bering on his early expeditions in the northern seas, exploring first for Peter the Great and later for Catherine. He says, "They cohabit after the manner of the human kind, the female below and the male above, and especially near evening time do they desire to indulge their passion. . . . He seizes her in his arms and indulges his passion with the greatest heat. During the coition he presses the female down and buries her in the sand by his weight so that only her head sticks out, and he himself digs into the sand with his front feet, so that he presses down and touches the female with his whole belly. . . . So absorbed are they and so forgetful of themselves that I could stand near them for more than a quarter of an hour without being observed. And I should not have been seen even then had I not struck the male a blow, whereupon with a great roar he attacked me so wrathfully that I got away with difficulty. But nevertheless when I gained an eminence from which I could look down he went on for another quarter of an hour with what he had begun."

At various times after giving birth to her pup, a cow will try to return to the sea, but her bull will quickly growl at her, and if she persists he will rush at her, inspect her, and drag her back into his fold. If she has been served, that is, if she is no longer in heat, he will let her go. She will slither off into the water, spend some time at the water's edge scrubbing off the dirt and grime of the beach with her flippers, and then gracefully swim out to sea to fish.

There is, in the fur seal, nothing equivalent to the formation of new groups such as we saw in the howlers. If a harem is considered a group, then this is a relatively fluid one, for the cows often change hands on the beach and the served cows leave while gravid ones keep hauling in during midsummer. If the whole mass of seals of the Pribilof Islands are considered the group, they do not split and seek new islands, which is the subject of Kipling's delightful story about them, but their homing instinct brings them back to the Pribilofs, and usually the same rookery, if not the same territory on the rookery. Originally the seals of the Commander Islands and of Robben Island must have been the same species as the Pribilof seals, but they have all gone to their same homes for so many years that now each has different physical characteristics and can be classed as separate species.

In the relations between seals, and their coordination as

a group, calls of a limited kind do play an important part. The bulls roar and bellow as a prelude to all their male activities and it is a note of authority, of challenge, demanding submission. Often the cows will bicker among themselves in the harem, and this never pleases the bull. He is the only one who has the right to fight so he gives a grunt and a growl or a whistle and a chuckle and the cows instantly stop their noisy chatter.

Calling is even more important between the young pups and their mothers, for not only do they find each other by this means but also they can to some extent recognize their actual kin (although smell is admittedly the most important and final clue to identity). The cow's call is something like that of a bleating sheep (only louder) and the pups give the bleat of a lamb. After feeding for five or ten days in the sea, a cow will immediately start calling the moment she hauls in. Sometimes a number of pups will bleat and come to her, but if they are not hers she will bite them and keep them off. Often only her pup will answer and he will come to relieve the pressure of her udder. But it is a casual affair for if the pup cannot reach her because of her position, she may let the pup go on bleating while she takes a nap. The pup may eventually get to her and if not she will change her position. Sometimes very young pups become separated from their mothers because the mothers have been carried off by an idle bull or the bull of some nearby harem. She cannot move for she is still in heat and is therefore highly coveted, so her pup must come to her to be fed. They will call back and forth to one another for long periods as the young pup struggles across the difficult terrain until they are finally reunited.

In the 1898 report there is an interesting description of a pup that was stillborn. The mother showed far more interest in this dead infant than was usual and tried very hard to make the limp object suckle. The cow became so excited that she began biting other cows and put the whole harem in a commotion. But the curious thing is that since the infant did not bleat, the mother never gave her call. I think this illustrates an important point concerning all the interactions of individuals in an animal society, namely that between the individuals there is a complex network of stimuli and responses, and here the stimulus for the mother's bleat was lacking, although the attention toward the pup was if anything greater. But in a few hours the cow had lost interest and left the pup to rot by her side and be eaten by the screaming Burgomaster gulls.

One strange fact is that there are no elaborate calls connected with danger. They do make a commotion if a man approaches them, but the actual warning comes usually from the alarm cries of the gulls upon which they seem to rely. They do not stand guard or post sentries but merely ignore the outside world until the birds give the signal. Seals have some natural enemies in the seas, particularly the killer whale, which can occasionally be seen rolling leisurely into the inlets of the Pribilofs. But the seals give no danger warnings; they do not rush out of the water or try to escape. This is one of the main reasons that sealers call the seals stupid, for not only do they show no fear of the killer whale, but they let man come in, year after year, to drive off droves and droves of bachelors from the hauling grounds, yet they never learn even though many may see the killing. The constant harping on this matter of seal stupidity in the early literature may be partly a justification of killing, for to the public they will remain nice soft furry animals with great big brown eyes, but it is mainly the fact that this social existence of seals is so complex and so human that it is hard to get used to the fact that they differ from us in other ways. We have quite accepted the fact that sheep are stupid but we still expect seals to think the way we do.

When the pups are very young, there are two instincts which become evident in a matter of a few days and both of which dovetail into the eventual social urge of the seals. One is the urge to romp and play and the other is to form groups consisting solely of pups; these infantile aggregates are called pods. There is no doubt that this urge to group with animals of a similar age is a great advantage to the small pup. It overtakes him in a matter of days after birth and it serves to keep the young ones away from the thought-less crushing of the busy bulls. At first on land these small-fry gangs bite and nip at one another, wriggling about and squirming much like the puppies of a dog. They will be disturbed only when a cow comes crashing in among them to feed her babe. She will send all the others scooting to both sides, flop down, and let her pup suckle, until he is quite bloated with milk.

After about two weeks on land the pods of pups will be-gin to work toward the sea and soon they can be seen tak-ing their first dip. The head seems too large and heavy at first for it keeps sinking and they become frightened, but they try again, some more courageous than others. After a bit they can be seen making excursions ten feet or so, and then they scuttle back to shore. In a day or two they have all taught themselves to swim and then the real fun begins. They turn and swirl and dive at one another, they try to catch the one holding the seaweed to get it away from him, as the pups of a dog will pull and chase for a stick. They have found a free and fluid world to which they seem marvelously adapted and they seem to enjoy it hugely. By the time the end of the season has come they are quite ready for the fall migration.

This tendency to group with specific kinds of seals ex-tends beyond the pups, as we have already seen. The bache-lors keep together on the hauling grounds and the young virgin cows seek the company of other cows in the harems. The bulls specifically seek the cows and because there are not enough cows for all the bulls to have a harem, only the idle bulls are alone and isolated. They always hope to elope with some cows but they are often frustrated and spend much of the season (until the very end) at the fringe of the great crowd, yet without their desired mates.

As we have said before, this whole aggregation of seals is primarily associated with reproduction. This is much more clearly the case here than with the howling monkeys, for with seals the feeding is done strictly away from the group. Furthermore, as one examines the group itself, it is clear that the territories, the harems, are the center of the grouping activity, and from this radiate all the lesser groupings. Not only that, but when the cows come into heat in droves, and the artificially large harems break up into a smaller more stable pattern, and when the cries and the roars of the rookery are at their peak night and day, this is the pinnacle of their social season. As the summer moves on the beaches become progressively quieter, fewer and fewer cows come into heat, and finally the starved and emaciated bulls begin to forget their wives and their territories. They become docile and haul back in to feed again. There is then a major let-down of the barriers; the idle bulls and half bulls may mate with a few tardy cows and there is a great intermingling of bachelors, cows, bulls, and pups. The strict grouping has gone with the sex urge, and finally as the weather turns colder they will slowly separate and wend their long way to the south for the long solitary winter.

RED DEER

Societies of different mammals undoubtedly arose separately and independently many times in evolution, and there is every reason to believe that the societies of howling monkeys, fur seals, and now red deer are quite distinct in their origin. It is not surprising that we find differences in their social life, but more difficult to explain are the many striking similarities. No doubt some similarities are to be understood in the basic behavior patterns that all mammals possess, such as the attention of a mother toward her child and the play among the young, and we may obtain better insight into these inherent mammal similarities when we compare mammal societies with insect societies. Also some similarities, and for that matter many differences, are the consequence of the conditions of the environment: the climate, the terrain, the other animals, and all the other multitudinous aspects of their surroundings. For instance, a striking difference is seen in the timing of reproductive activity in the howlers and the fur seal. The seal gives birth to pups and mates in a remarkably short period of time, apparently associated with the short summer season and the restricted time which it can remain on land because of the adverse climate of the winter. On the other hand in howlers both childbirth and coition occur at any time of the year in the almost uniform climate of the tropics. And there are other major differences which have their effect on their social existence: the howler lives almost entirely in the trees, the seal spends most of the year in the sea, and now we come to a strictly land form.

The special reason for choosing the red deer is that, like Carpenter with the howling monkeys, the British naturalist

F. Fraser Darling spent two years living with the red deer near the northwest coast of Scotland, opposite the Hebrides. He went there specifically to study their social behavior, and although his book is a careful scientific analysis, it is also fascinating to read, for each page reveals his ardent love of the highlands and his pride in his own knowledge and prowess in the very difficult art of deer stalking.

What Darling calls his beat is an area of about eighty square miles, fringed at the northern edge by the North Sea. It is a wild country with the Torridonian mountains rising three thousand feet; with lakes and with deep gullies or corries scoured by early glaciers. The vegetation is very sparse on the higher peaks, but in the corries and the other lower areas there are grass, heather, and many other scrubby plants. In a few areas there are woods, pine woods sometimes, and birches and alders in others. I will spare you the names of all the different places, each place with its own special beauty, for the names are all in Gaelic and it takes a little practice to tell one from another.

The weather is very changeable and almost always humid. The winds blow often, and the air has a purity and a delicacy familiar to all who have climbed a mountain or camped out in northern regions. The mean temperature is about 38°F in January to 56°F in July, but the maximum may on a hot summer's day go up to the 90's and down to 20° or slightly less in winter. There is much rain, which comes in hard puffs, but then it will clear quickly. In winter the snow comes frequently but usually does not stay long.

Stalking deer requires great skill because the deer are incredibly acute in hearing, sight, and smell. If the scent is carried by the wind the deer can smell man a half-mile away. Their hearing is almost as sharp; on a still day any slight noise can be detected at great distances. Since much stalking is done crawling on one's belly it is very important to wear the right clothes, preferably Harris tweeds, for any harder material will make too much noise scraping along the rocks and heather. The eyes of the deer are especially quick at picking up movements at a distance. If one remains stock still and the deer happen to be approaching, they will not come closer than sixty or seventy yards before they become frightened by the movements of one's eyes. The acuteness of their vision is even more strongly emphasized if it is compared to man's. Sitting near the crest of a hill and looking way down and across the corrie below with a 20 × telescope, an experienced stalker will find the deer and be able to count them and see if there are any worthy stags. But a novice taking the glass may see nothing for a long time because their color blends so easily with the vegetation. Any great movement by the stalker will disturb the deer, even though he can barely see them.

Red deer are large, not so large as our elk, but a good deal bigger than our Virginia deer (Plate IV). They need a lot of nourishment and the grazing grounds in western Scotland appear just barely sufficient to maintain them. But competition for food is not the only factor which limits the population. There appear to be certain psychological factors which prevent great crowding. For instance, if sheep are introduced into a certain area they will consume only a small part of the deer's food, yet the deer become exceptionally sparse in that area, so much so that much of the land is left ungrazed. Probably the continual presence of man and his collie dog do much more to thin the area of deer than the sheep themselves. But even when left alone in areas of abundant food the deer may reach an equilibrium that falls short of the maximum food supply. Darling finds about a total of 1300 deer in his census of the 80 square miles, about 1 deer to 40 acres. In the most suitable areas the deer may be 1 to 25 or 30 acres, and in other areas the density is less, but this average is maintained and curiously enough it is close to the equilibrium density of many other deer in totally different regions and countries.

The actual loss to the population in Darling's beat is

accounted for in part by man who shoots on the average fifty-five stags annually, but infant mortality is also considerable and only fifty per cent of the calves live through the first year. Foxes, golden eagles, and wild cats take some; the bad weather of October accounts for more. Also in October the hinds have difficulty finding good grazing, and often the milk, which should keep on right through the winter and the spring, dries up or becomes thin and the calf can no longer withstand the weather or the ravages of parasites.

In the western highlands calcium and phosphorus are not readily available to the deer in any abundance. This is severely felt since the stags need a great supply for their annual growth of new antlers, and the hind needs it for her milk. As a result they are avid for burnt ground and will munch the charred remains of plants, which are rich in salts. Even more striking is the fact that they will eat shed antlers, starting at the tips and working down, and Darling once saw a hind eating the antlers still on the head of a stag. If a deer dies, his carcass will be devoured by gulls, crows, ravens or an eagle, and then the maggots of insects will finish the job so that nothing but the white bones are left. Then the deer come and eat the bones, for they urgently require the calcium phosphate.

There are definite territories associated with feeding, and each deer belongs to a specific and clearly defined region. These regions are different in summer and winter, for in the summer they move to the upper slopes and in the winter or during unfavorable weather they come down to the valleys below. The lower winter territory is a fairly circumscribed area; it is sharply bounded on three lower sides, the fourth boundary being the road to the summer territory and the mountain tops. In summer there may be a mingling and crossing of the groups but this is only temporary and each returns to its own terrain eventually. Within these major territories the hinds and the stags keep apart except during the short rutting period in the fall. So there are hind groups and stag groups, as well as harem groups during the rut, and as we shall see shortly, each of these has its special characteristics, its special type of integration. The only animals that never group at any season are wounded deer and a few old stags decayed with age.

The extent of the summer and winter ranges are undoubtedly governed primarily by food. Each herd has its pasture lands. There are no fights or violent defense even of the rigidly confining winter territories; rather the animals seem to stay within boundaries. Darling did an experiment which showed this point most clearly: he placed corn out every evening at a certain place and finally, after waiting a month, some deer began to eat his corn. Then each night, as they grew bolder, he would place the corn in different spots and they would find it. But when he tried to lead them across a stream which they could easily ford (but which was the limit of their territory), they refused. The corn was no longer in their land.

As in the seal, seasonal sexual activity plays a powerful role in their social existence. Red deer differ, however, in that not only do they come together during the mating season, but they herd at all other times of the year as well.

In preparation for the fall rutting season the stags grow a new pair of antlers. They cast the old ones in spring and all summer the new ones, encased in soft velvet, burgeon upwards and branch into the beautiful full grown antlers. These great deposits of calcium are all laid down in a few summer months and then the stag scratches and rubs the velvet off so that the hard surface is exposed.

These antlers and their growth are controlled by the male sex hormones. If a fully developed stag is castrated he will shed his antlers, and no more antlers will grow, or if they do they will be deformed. But the growing of antlers is a costly drain on the energies of the stag; it would be much easier to keep them and add to them each year, as a sheep does with its horns. Even more curious—why do they exist at all and of what use are they to the stag? It has been thought that they are involved in sexual selection, and that they serve both as a weapon to keep away other stags and a powerful attraction to the hinds. But Darling shows that this is hardly certain for there is among Scottish red deer a frequent occurrence of stags without antlers, called "hummel" stags, which are occasionally very successful as harem masters. They may fight with great energy and success and they keep many hinds. There is little chance, however, that such hummel stags will become abundant, for the owners of the forests kill them as soon as they find them for the same reason that a husbandryman will kill a deformed bull. There are other possible uses of antlers but none which seem even faintly reasonable, so here is a huge structure whose use remains uncertain and enigmatic. Maybe we simply have not hit upon the secret yet, or it may be that it simply has no important use, and is just there.

After the antlers are hard, other physical preparations for the rut take place. The stag's testes begin to swell and enlarge, and the manufacture of sperm begins. Much experimental work has been done on the cause of this phenomenon in other seasonal forms, and it is now known that the comparative lengths of day and night rather than the summer temperature start these processes off, the same mechanism in fact which brings flowers to a plant. Also the whole neck of the stag becomes greatly enlarged, including the thickening of his mane. His larynx swells so that he can now produce a roaring sound that is the clearest sign that the rut is in progress. It is quite possible that these features are all part of the bluffing mechanism which we have already talked about. The stag like the bull seal or the male howler puts on a great show, and like the seal he actually fights, although the fights are not so bloody and there is more show than fighting.

After their antlers are shed and while they are growing

new ones the stags form groups which last until the rut approaches. Occasionally during this period and especially during the rut some stags will wander far away into other territories, sometimes as far as seventy-five miles away. The stags are very docile and friendly when in velvet and appear to take easily to one another's company. When the velvet is shed the flocks seem to get even bigger, and they may make occasional digs at one another but it seems only a mild experimental irritation.

When the rut finally begins, toward the end of September, the stags separate and are constantly on the move, in a short, determined trot. The trigger that sets them off is the fact that the hinds have come into heat. The older and more aggressive stags literally round up a number of hinds, as many as they can get and hold together; sometimes more than fifty, and often as few as a half dozen. The hinds have with them their young, immature offspring of the previous years, and they appear to submit passively and unconcernedly to the irritable and nervous herding of the stag. He will trot round and round them almost continuously, first in one direction, then in the other, only occasionally lying down for a half minute and then up again; sometimes roaring as he runs with his snout outstretched. One of the most curious observations of Darling is that the stags masturbate often several times within an hour as they are tending their harem. The antlers are apparently an erotic zone, and by gently rubbing them in the grass the stag has a quick erection and an ejaculation, the whole lasting but ten or fifteen seconds.

Actual copulation was observed only twice by Darling in his two years of observation; apparently it must take place primarily at night. On one occasion the stag tried to mount a hind twelve times, and only on the last try when his hind feet left the ground was it certain they had copulated. In the other instance the female showed much more interest and licked and rubbed the stag before he served her, which he did and then soon after she enticed him again. After the second copulation the stag lay down to rest for an unusually long time.

Should another stag approach a harem, and if he feels he is no match for the harem master, he will gallop off as the master approaches, head low and roaring. But if they are equally matched they may fight. They first run at each other, heads down, but they stop when the antlers clash, putting little force behind them. This is just playing for position. Then suddenly one will lunge at the side of the other trying to give an up-butt under the belly. If the opponent is quick enough he jumps to one side and meets antlers with antlers. If he is luckless, however, he gets banged, bruised and cut on his underside and soon retires defeated. A stag does not eat during the period he keeps a harem and so in two weeks time he is well nigh exhausted and can be easily removed and defeated by a fresher and often younger stag. The spent stag will wander up to higher ground to feed and recuperate, and may even come down again after he is fresh to try his luck once more.

The next June or July, just before they calve, the hinds break away from the hind groups, followed by their young of the two or three previous years. They are so eager for solitude that they try to chase away their own young. The two- and three-year-olds will stand off at a distance, but the yearlings consider her aggressive behavior a part of a game and do little else than prance about and come right back. Just before parturition the hind will be extremely restless, she will alternately lie down, rise, and pace about. Finally in a standing position she drops her calf, and in an instance that Darling was able to observe in detail the hind paid remarkably little attention to the newborn for some while, but merely stared at it. Finally she licked it a bit but soon she wandered off about a hundred yards and began to graze. The calf could not stand yet, and the mother returned a couple of times in the first day to suckle it, but she seemed to do this very casually.

In a few days, when the calf is on its feet, the hind suddenly seems to blossom as a mother. The calf now suckles almost constantly and the hind nuzzles and licks her calf, especially the ears. Of this Darling says, "If one hind starts doing it, the others will soon follow suit, and it is very amusing to see several calves standing like little boys having their ears washed by their mothers." It is probable that the hind has little milk for the first three days and that the change in the mothers' attitude is influenced by the coming in of the milk. But it is more than the pressure of the udder, for it is now a well-known fact that the hormones in the blood which stimulate milk production also stimulate maternal behavior.

Darling reported one case of the reaction to a stillborn calf which is interesting when compared to the case of the seal's dead pup that created such a commotion. He found a group of three hinds circling the calf and soon the whole herd of sixty or seventy hinds was centered about the calf. They did not come very close to it and soon they lost interest and moved off. They are very casual at first with a live calf, but a dead one does appear to disturb them in some way.

The most interesting aspect of red deer, from our point of view, is the coordination of the group, and most especially the hind groups. There is a rigid integrated pattern that seems to help them in preservation from enemies. The whole herd of say forty to fifty hinds may be separated into families during the calving time, as we have just seen, and thus each family has a leader, the mother hind. She is ever on the alert for a sight, a smell, or a sound of danger, with her head held up high, and should it come, she gives a signal, a warning bark. But her followers do not, as one would expect, stampede off at the warning. Instead they fix their eye on the stalker, or whatever is the cause of disturbance, and then move off in an orderly fashion, always keeping the enemy in sight. Confusion comes only if a hind smells the stalker but cannot see him; she appears very uneasy and disturbed.

After calving, the hinds come together again in a large herd, and then all the previous minor leaders assume a passive role and there is one leader for the whole group. This prime sentinel is usually an older hind, and almost invariably has a calf with her. She will retain this position of leader as long as she remains fertile, but should she become barren (a "yeld" hind) then she will recede to the ranks. Apparently this instinct to help preserve others is connected with the maternal instinct to watch over her young. There are, as we just said, instincts stimulated or enhanced by the hormones that are secreted into her blood when she is with milk. Integration is therefore again closely connected with reproduction, and this is emphasized by the fact that stag groups are not nearly so well integrated. It is for this reason that Darling calls the society of red deer a matriarchal rather than a patriarchal one. Only in the rutting season does the stag dominate, and then solely for the purpose of satisfying his egocentric sex desires. Even during the rut the hind leader continues to play her sentinel role, and protects not only the other hinds and calves but the roaring stag who is too preoccupied with his urges to keep much of an eye out for any danger other than wife-stealing.

Besides the leader there is a second-in-command who occupies the opposite end of the herd, the rear guard. A beautiful example of how they operate together occurs when the hinds must go down into a little gully and up again before they can escape the observer. Since they are in continual fear of ever losing sight of the enemy, the spindle-shaped herd (for they move in a perfect spindle with the leader at one end and the sub-leader at the other) will move down into the gully, but the rear leader will stop and stand motionless staring at the observer while the rest leave her and go from the observer's sight. She remains there with eye fixed until the first leader is up the other side of the gully and stops and fixes her eye on the man, and then the rear leader trots off and again takes her proper position. This orderly retreat is lacking only during a stampede, which may be caused by the approach of particularly irritating and hard-biting flies. The hinds will go into great commotion, and tear off in a fan-shaped group with no care or thought, in a frenzied rush to escape the insects.

In yearlings a difference between the young stags and hinds in their sentinel work is already evident. From the beginning the young hind will learn to look out for herself, but the little "staggie" pays no attention whatsoever and depends entirely on the females for danger signals. This is true until he leaves the hinds at three years and then he begins to watch for himself.

But stag groups do not have any leaders or any organization. Any animal may give the alarm and they will leave with no consistent grouping, sometimes in single file and sometimes they split and go in different directions.

There is some tendency among deer, as there is in seals, for the age groups to stick together, and it is not uncommon to find groups of three-year-old stags or three-year-old hinds, or childless hinds. This, however, is only a tendency and not an invariable rule, for often one will find, for instance, a young staggie with an old stag of ten or twelve points.

One other aspect of this grouping is interesting: when a group of hinds or stags are down and resting they arrange themselves so that they face in all the different directions, to increase the possibility of detecting an approaching enemy. If a single deer sits, he faces downwind so that he may use his nose for that which is behind him, and his eyes for that in front.

The importance of the voice of the deer as a means of communication has already been abundantly illustrated.

There is the mating roar, which can be produced by the stag only during the rut when his larynx temporarily enlarges; there is the warning bark of the hind in the presence of danger; and there are the grunts and the whines that bind mother and child when the young calves are about.

Finally I should like to say that the calves, like most mammals, are very active at play and this no doubt has an important relation to their later social existence. Darling has studied their play carefully and finds that they have four games: racing, mock fights, tag, and king-o'-the-castle. The racing consists of a sort of excited running together. For the mock fights they pretend to be angry and stand on their hind feet as though to hit one another with their fore feet, but they never do. The tag is not very systematic and no one calf is "it" but they all rush about in different directions trying to tag one another with their fore feet. Sometimes the tag is played around a hillock and they go round and round, only to be surprised by one of them doubling back and meeting the other one head on. There are certain hillocks that seem to be special playgrounds for king-o'-the-castle. A calf will run to the top and rear and at this bidding the other calves will scamper off from their mothers and try to get to the top of the hillock too. And when they have played until they are all breathless they will stop as quickly as they started and trot back to their mothers.

In his book Darling makes an interesting comparison of the social instincts of domestic sheep with those of the red deer. Certain breeds of sheep, such as Mountain Blackface used so much in Scotland, have no instinct to flock and will tend to separate and spread so that they will cover wide areas of the mountains of thin vegetation. Presumably this tendency to range has been bred into the sheep by years of selection on the part of the Scottish shepherds, just as Spanish shepherds have bred the Merino sheep to flock. In Spain the sheep have to make a long migration from summer to winter grazing grounds and back, and there is a great advantage if they keep close together and can be easily herded by a dog. By breeding under domestication, then, it is possible to accentuate or to suppress certain social instincts, and the lesson we may learn from this is that these instincts are to a marked degree controlled or affected by the heredity of the animal. The tendency to flock or to form rigid groups as in the hind groups of the red deer is not solely a habit acquired by learning on the part of the younger deer, as are so many of our human customs, but it is innate in each individual; it is a part of his inheritance. There is certainly some learning, as there are some inherited factors in our own human social existences, but in the deer the importance of the innate instinctive action is relatively greater and no doubt this is one of the reasons the red deer society is so rigid and inflexible compared to our own.

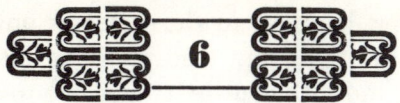

6

BEAVERS

IT WOULD not seem quite right, in discussing mammal societies, to neglect the famous beaver. They differ from our previous examples in that their groupings are relatively small family groupings, and also, despite their fame, surprisingly little is known about the beaver. We know his works, and they are indeed remarkable, but so much of his life cycle, especially the breeding and the rearing of the young, takes place hidden in his lodge, that they cannot be observed in any detail. On the other hand, the fact that his social unit is that of a family, and the fact that he produces such remarkable feats of engineering, does anticipate conditions we will find in the social insects; and for that reason the beaver will serve as an excellent bridge to our discussion of those more primitive organisms.

The beaver, including both the present living species and the prehistoric forms, some of which were very much larger, has literally been responsible for changing the face of the earth. Their stubborn and persistent instinct to build dams and block up streams has caused the gradual appearance of meadows, deposits of rich and fertile silt, the loss of woodlands, the production of peat bogs and fens, landmarks which exist today over most of North America, Europe, and Asia. The animals in this country are now somewhat restricted as compared to their previous widespread distribution; they are found mainly in the northern half of our North American continent and in regions of high altitude. In Europe the species has mainly disappeared, although they existed in the Po valley in Italy as late as the sixteenth century and there are references to beavers in Wales in the twelfth century.

The beaver is happiest in the water or under the cover of ground or the thick and sturdy roof of his lodge. He is generally shy, and although he comes out to feed (usually at night) along the banks and even some distance in the woods, the slightest disturbance will send him back into the water, to the underwater tunnels of his lodge, and finally to the inner chamber where he is safe. The lodge usually consists of a great conical mound of sticks, saplings, and mud. All the entranceways lead under the water and in the center there is one fairly large chamber carpeted with grass and shredded wood. In order to keep this lodge safe from intrusion the water level of the pond must rise above the openings, and a certain depth of water is even necessary for safe swimming. These problems are solved by their instinct to build dams, for in this way they manage to keep the water level high, or even make some new location habitable where previously it was not.

Most creatures accept the environment as it is and during the course of evolution their habits and their body structure have changed so that they become adapted. But beavers, like men, change the environment to suit their way of life, and instinctively they mold it to their liking.

The beaver, like the fur seal, has a valuable coat of fur and has also some specific body features that make him an agile swimmer (Plate V). With his webbed hind feet he pushes his way through the water; his flat broad leathery tail is mainly used as a rudder and only upon occasion for an extra push. This peculiar, highly muscular tail can be used for other purposes: sometimes as a seat, sometimes to carry mud by clasping it between tail and body, and sometimes to slap the water as an alarm signal. The forefeet are not webbed but have claws which can dig, almost like hands. One of the most striking features of a beaver is his powerful incisor teeth, with which he can gnaw through and fell trees more than a foot thick.

A beaver's first impulse upon entering a new area is to

chew off sticks and even sections of saplings and thrust one end down into the bottom of the stream. By continuing this process with a perseverance that has made the beaver a symbol of industry, they manage to get a network of small boughs across the stream. Now they begin, from the upstream side, to add mud by scooping it up from the bottom and plastering it along the basket work. Soon more branches will be brought on and then more mud, and finally the silt and the debris of the stream begin to help the beaver and clog up the holes so that finally it holds tight, and then as the water rises it needs new work at the edges, or perhaps at the top. Dam-making is an activity which is never finished, for the dam needs constant attention, either because of damage or sometimes because the pond silts up and in order to obtain the desirable depth the dam must be higher. Dam-building and lodge-construction, which occur primarily at night, are most active as the winter approaches; little work can be done when the pond freezes over and therefore the dam is put into good repair and even the lodge will get an extra coat of mud.

A dam of one-season's building is not likely to be very large, but in a big pond where numerous families have lived for some time the dam may attain considerable size, an accumulation of years of activity. A beaver dam may reach a length of five hundred feet and be more than ten feet deep at some places, although the usual dam rarely exceeds five feet in depth. There is a great deal of folklore connected with the dams of the beaver and we are asked to believe that the shapes of their dams are as perfect as an engineer would have it, but the naturalists assure us that though they occasionally construct a dam with a scientific upstream bow like man-made dams, this is pure chance, and just as many curve downstream or are serpentine. The beaver apparently follows from one bit of construction to the next, and the result is remarkable enough without having to endow the beaver with a profound knowledge of hydraulic engineering.

Some dams make ponds that are not inhabited at all, and these are apparently built to provide transportation. It is far easier for the beaver to float his branches and logs and pull them along with his teeth as he swims than drag them on the ground. Nevertheless there are often trails or tote roads down to the pond where the beavers have worn the ground bare by dragging timber. But the beaver is so much more pleased with water transport that if he does not build a dam and a pond for that purpose, he may build a canal, digging it out of the ground with his forepaws. These canals are sometimes remarkably large; a good record is a canal in the Adirondacks which is 654 feet long and only 15 to 18 inches wide, although as a rule they run less than 100 feet. When canals are built in sloping ground they have the added feature of having locks (i.e. small dams) to keep the water level up in all its parts, substantiating the point of view of Lewis H. Morgan (the first naturalist to make a really comprehensive study of the beaver) who believed that the canals exceed even the dams in their engineering ingenuity.

In the summertime the beaver eats wild rose or raspberry bushes, different roots and grasses, thistles, and a variety of other plants. But in the winter his diet is confined pretty much to the bark of trees, especially the aspen and the willow. He may eat all the outer bark of a young shoot but with older parts he discards the hard outer layer and eats the green tender covering of the wood.

In late summer or autumn the beaver will select a tree and gnaw at the base with his great incisors. His teeth are an efficient device for with two bites he will be through a bush a half-inch thick, a four-inch aspen will take about an hour, and a tree eight to ten inches thick can be cut through in a night. Sometimes on the larger trees they work intermittently, and it may be the work of more than one beaver. They do not, as is sometimes claimed, control the direction of the falling of the tree, but once the tree is down they may then further cut it up into sections that can be transported readily. With these they make off, down the trails, down the canals, finally to their own pond and then they force the sticks and branches down into the mud in the bottom so that as winter approaches they will have a great heap of water-soaked wood at the bottom of the pond. This is the winter store, and when the pond freezes over they can go under the ice from the underwater exits of the lodge and easily reach their food supply.

A lodge is occupied by one family only, but a family may be quite large and often consists of about eight individuals. These include the father, a mature male, his wife (for they are monogamous), and various-sized offspring. The latter are kept in the parental lodge for a couple of years and then they are forcibly evicted, chased away to fend for themselves and seek mates and start their own lodges. They may do this in the same pond or they may move elsewhere if the population of the pond is already great for its size. In some large ponds there may be as many as eight lodges, although the more usual figure is three.

Nothing is known of the rut, which probably occurs in mid-winter, because it all takes place inside the lodge, hidden from us by the snow and ice, the sticks and the mud. The offspring (usually two to six in number) are born after about three months' pregnancy in the early spring, and in about six weeks to two months they are weaned and may sometimes be seen sunning themselves on the top of the lodge. Like the young of all mammals they are very playful and romp about in the water and on land, nipping and tugging and rolling about.

During the period of parturition and early infancy the male apparently leaves the lodge and lives in burrows in the bank of the pond, another of the beaver's feats of engineering. These burrows may be quite long and stretch far into the bank, sometimes as much as thirty or forty feet. In some cases the whole family may live in a burrow rather than build a lodge.

There are various ways in which beavers are coordinated in their society; the famous North American naturalist Ernest Thompson Seton has given us a vivid account of their language: "Being essentially sociable, the beaver has many methods of communicating ideas. The first of these is vocal. Young beavers wail like a crying child. Older ones hiss in menace or utter a querulous 'churr.' When two meet in the pond I have several times seen them nibble each other's cheek, at the same time uttering a chattering noise; I suppose it is a friendly salutation.

"Another important means is the splash signal. While watching beavers at Yancey's [in Yellowstone Park] I learned that at once, on discovering danger, each beaver gives a great slap with its tail and dives; this is understood by all and repeated by all as they dive. . . . The sound of it on a quiet evening is very far-reaching; it is in fact two sounds, one a loud slap as of a paddle, followed at once by a deep hollow plunge, as though a ten-ton boulder had been dropped in the water.

"But there is another way which may be called the mud-pie telephone" and Seton goes on to describe the secretion of the glands near the anus called "castor" or "castoreum." This is deposited on the surface of a small heap of mud and it is in every sense a calling card. Any other beaver who passes by will know the sex and probably other personal details of the caller, just by smelling the castoreum, much as dogs use urine as a source of information. It is presumably used by males to intimidate and frighten off other males from their territory and also it is used by a mature male to entice a mate into his domain; he will put his advertisement on every prominent point for miles about. Sometimes when older beavers romp and play, which they do upon occasion, they will surround the playground with deposits of castoreum and, encompassed by its heavy musky odor, indulge in their orgy of play.

In many ways the greatest degree of cooperation is seen in the physical structures they build: the dams, canals, burrows, and lodges. These are all cooperative and mutually beneficial ventures. They are, so to speak, bound together by their works; this is a communal activity and one which facilitates their feeding and their reproduction and even protects them from their enemies.

SOCIAL INSECTS

BEFORE entering into the details of any particular insect society it may be well first to say something about insect societies in general. Insects, after all, are very different from mammals and therefore it will not be surprising to find that their societies differ in many ways from mammal societies. Nevertheless the three functions which we are following through the organic world—taking in energy, reproduction, and coordination—not only can be seen in insect societies, but they stand out even more clearly and more boldly. This is primarily because insect societies have to a high degree a fixed and inflexible integration, a rigid pattern, and this rigidity makes their social activities all the more sharply defined.

Like mammals, insects have formed societies a number of separate times in evolution; William Morton Wheeler, the great entomologist and authority on insect societies, estimates that they have done this no less than twenty-four times. Furthermore it is known from the study of fossil insects which are beautifully embedded in amber (the fossil resin of early coniferous trees) that these insect societies have existed for many millions of years. This, of course, is also true of the many solitary forms, but it shows that their social existence has a long record of stability.

On the whole, insects are much smaller than mammals, and their nervous systems are also more rudimentary. They do not seem to be capable of any great mental deductions, but they do respond in specific ways to a large number of external stimuli. And also their senses are remarkably acute. In some the eyes are well developed—for example they are important in the social life of the honey bee—but in others, such as the army ants, the eyes have virtually disappeared and they are blind. Some insects have keen hearing; the chirp of a cricket is not solely to cheer or annoy human beings but is primarily to call a mate who finds the shrill note quite irresistible. But most remarkable to us, for we are comparatively insensitive, is their sense of smell or what might more properly be called their chemical sense, for it sometimes involves actual contact with the chemical rather than solely the recording of its vapors. A small piece of a female cockroach will send a group of males into a wild ecstasy of anticipation, and the odor of a particular flower on a worker will help guide the other bees to the source of nectar. In fact the whole galaxy of scents in flowers has presumably arisen in the course of evolution as a mechanism to entice insects who serve to pollinate them. And all these senses, as we shall see, play a decisive role in the integration of the insect society.

The very first step toward a social existence in insects, according to Wheeler, is the prolonging of adult life, and this, he conjectures, is a result of the specialized feeding habits acquired by certain insects to avoid competition with others. This is especially obvious in the case of vegetarians where the food may not be very nutritious, as is the case in wood-eating termites, or not very abundant as in dung beetles, and in both cases time and perseverance are required for feeding. The immediate effect of this prolonged adult life is that the young will be in contact with their parents for longer periods of time, leading ultimately to a dependency of the young upon the parents, perhaps by loss of some self-sustaining ability on the part of the growing larvae. The most primitive condition of a true social existence exists when the adults take care of their young by providing food for them, and perhaps building a nest for them, but even this nursing may be graded. For instance among some of the social wasps the mother leaves enough food to last the whole of the offspring's development, and in other cases she will keep bringing food as they grow and therefore prolong her period of attention. The next step toward a more elaborate society (for instance among honey bees) is the production of neuter offspring called "workers" (actually nonreproductive females), and these then help in obtaining food and feeding the larvae. They become efficient nurses, allowing the mother to concentrate on egg-laying. The fertile males of many social insects, especially the social wasps, the ants, and the bees become quite superfluous except for fertilizing the female, an act which is quickly accomplished and usually only once during the lifetime of the female. She stores the sperm in a special sac and uses them sparingly on the eggs as they pass out of her. The male then has no part in the well-developed colony once the queen is fertilized. Wheeler says of him that ". . . he has not even the status of a loafing policeman, and all the activities of the community are carried on by the females, and mostly by widows, debutantes and spinsters. The facts certainly compel even those who, like myself, are neither feminists nor vegetarians, to confess that the whole trend of evolution in the most interesting of social insects is towards an ever increasing matriarchy, or gynarchy and vegetarianism."

The sorts of food for which they have acquired a taste are varied, and some, such as the more primitive social wasps and some ants (e.g. the army ant), are carnivores and eat insects and often the flesh of other animals. Others, for instance the honey bee, eat the nectar of flowers; others such as numerous ants eat plant juices. Often these juices have been sucked up by plant lice, which the ants keep as domestic animals, and by stroking them the ants cause the plant lice to regurgitate some of the juice which the ants avidly take from them. The honey ants take the plant juices and store them underground; certain of the workers hang from the ceiling of small chambers and become living vats; their abdomens are like round marbles distended with

PLATE I. Two young howling monkeys in captivity. (Photo by C. R. Carpenter)

Plate II. Fur seal cows. (Photo by Karl W. Kenyon)

PLATE III. A bull fur seal with his harem. (Photo by Karl W. Kenyon)

PLATE IV. Red deer in captivity. A stag, three hinds, and two calves. (New York Zoological Society)

PLATE V. A beaver sitting on his tail. (Copyright Walt Disney Productions)

PLATE VI. *Above*: An army ant queen swollen and laden with eggs. She is actually
labor and the surrounding workers are snatching the eggs. *Below*: A column of marc
army ants. This column may extend more than 300 yards in length. The queen is
above the center of the photograph and surrounded by some excited workers. (Ph
by T. C. Schneirla)

VII. An army ant bivouac. The ants cluster and hook to one another in festoons in this way pass the night during their nomadic phase. (Photo by T. C. Schneirla)

PLATE VIII. A termite nest in the Suk country in Africa.
(American Museum of Natural History)

honey which they will regurgitate when they are stroked (Fig. 1). The harvesting ants eat grain which they store in their underground granaries; carpenter ants and termites eat wood and obtain their energy from the cellulose. But most remarkable of all are the ants which keep fungus gardens; they are true horticulturists.

One of the most interesting aspects of feeding in the so-

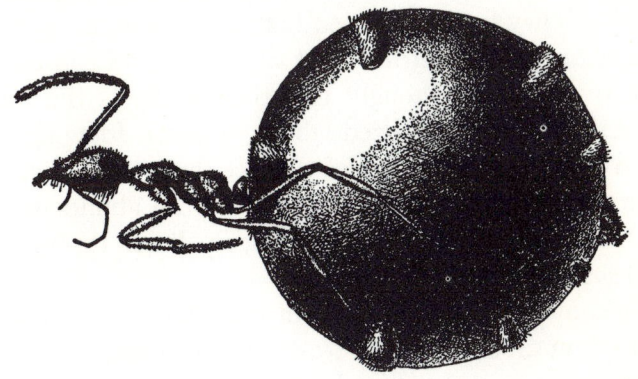

FIG. 1. A honey ant storage worker. (From W. M. Wheeler)

cial insects is the tendency to share or exchange food or salivary secretions. This may take place either between adults or between adults and larvae. It is a process which appears to serve a number of functions: one of binding and coordinating the individuals, and one of feeding those members of the colony that do not or cannot go out and get food for themselves. The relatively helpless and immobile larvae or the egg-laying queen may be given food by this food-exchanging process. This is an example of division of labor involving a sharing of the food.

To illustrate the extent of this food sharing, recently some biologists in England have done an interesting experiment with honey bees. From a hive of 24,500, they fed six worker bees radioactive phosphorus, a substance that can be detected in very small quantities because of its radioactivity. These six bees were then allowed to return to their hive and after 27 hours each bee in the hive was tested to see if it possessed some of the phosphorus. It was found that 76 per cent of the actively foraging workers had some, and it could be detected in 43 to 60 per cent of all the bees. After two days all the older larvae that were in open cells were radioactive. By this ingenious test it was possible to show that the food exchange is a rapid and busy activity in a beehive. They further showed that this is not only a matter of distributing the nutrition but also a means of communication. It used to be thought that the activity of the various workers in their specialized labors (nursing, foraging, etc.) was a function of the age of the bee, but now it has been demonstrated that it is a function of the abundance and kind of food which they presumably learn through this food-sharing telephone system.

To come now to the matter of reproduction we find that (with a number of notable exceptions) higher insect colonies are not only matriarchal but also monogamous. This is not always true for in the termites the male or king continues to live with the queen and repeatedly fertilizes her, and there are also some cases of polygamy where a number of queens live in a colony. But in any event it would be fair to say that even though the colonies may achieve a tremendous size and consist of thousands of individuals, they are in most cases but a family, one mother with her vast offspring. These offspring may be of many different sorts: they may be new males and females that go elsewhere to start new colonies or they may be nonreproductive females. These sterile females furthermore are graded into almost a continuous series of sizes in some cases, and in others they may fall into distinct groups, often soldiers and workers, or soldiers and major and minor workers (Fig. 2). Each of these castes will to some extent divide the labor and most curious of all, they exist in fairly fixed proportions even though they are constantly dying and being replaced by new ones, all from the fertile queen.

The problem of how these castes are produced is one which has interested biologists for many years, and a good deal of the story is now well known. The difference be-

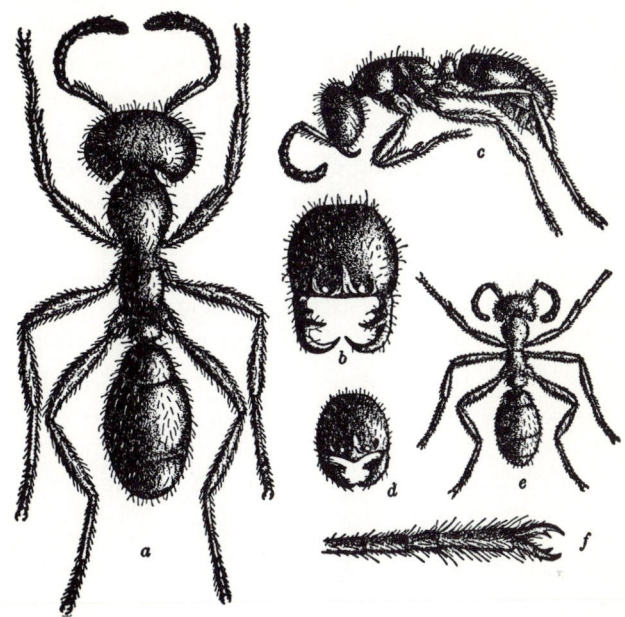

FIG. 2. Different sizes of workers of one species of ant: *a*, soldier; *b*, head of the same from above; *c*, medium sized worker; *d*, head of same; *e*, minor worker; *f*, end of a leg to show the claws. (From W. M. Wheeler)

tween males and females is a genetic one (as it is with most animals and plants) but in this case the males are produced from unfertilized eggs while the females come from fertilized ones. The females then have twice the number of chromosomes (the bodies in the nuclei of the cells which carry the heredity factors or genes). One set comes from the father and one from the mother. The males, on the other hand, have chromosomes only from the mother and this difference somehow creates all the structural differences so apparent between the sexes. The neuter castes also come from fertile eggs and therefore are fundamentally female, but how is it that they differ from one another or differ from the queen? The situation is perhaps best known in the honey bee where there is some evidence that the presumptive queen's large egg cell not only receives a greater quantity of food than a worker's cell, but the queen gets food which is richer, a secretion from the pharyngeal glands of the worker which is called the "royal jelly." The difference then is thought to be one of nutrition for the eggs are the same, but the quantity and/or the quality of the queen's diet determines her fate. The workers are incapable of reproduction because they have been subjected to what Wheeler calls nutricial castration. The same situation is believed now to exist in other forms such as ants and termites, although the evidence is by no means clear as yet. It is thought that in this way all the different-sized workers are produced, even though the different sizes may have radically different proportions. This is seen especially vividly in the soldiers of ants and termites, where the jaws of these larger forms seem disproportionately big and fierce.

In his *Origin of Species* Darwin worried specifically about the problem of the evolution of social insects because there seemed to be so many individuals that were sterile, and since his whole scheme depended on sexual propagation he says that he thought at one time that this might raise serious doubts concerning his theory of natural selection. For after all how is it possible for these neuters to transmit their selected variations since they are sterile? But then he points out that selection cannot act here solely on the individual, but must act also on the whole society, on the whole family. If the queen produces offspring which instinctively perform certain tasks effectively, more so than a competing colony, the daughter queens and their new colonies will have an advantage over any others less well-adapted. But the unit of selection here is primarily a single social unit, and one of the ways in which the insect society is kept cohesive is by the limitation of the reproductives to a single pair (or at least to a small number), for on account of this all the existing facets and all the future changes in their intricate society must funnel through the hereditary make-up, the germ-plasm of that pair.

Considering all this, it is not surprising to find that a queen, for example a queen ant, will have all the instincts of the colony. This is shown when, after her nuptial flight (and the last of the male as far as the fate of the colony is concerned), the newly fertilized queen will set out to start a new colony. She first removes her wings and then digs a small hole in the ground and there closes herself off in a chamber (Fig. 3). Here, after some time, she lays her eggs,

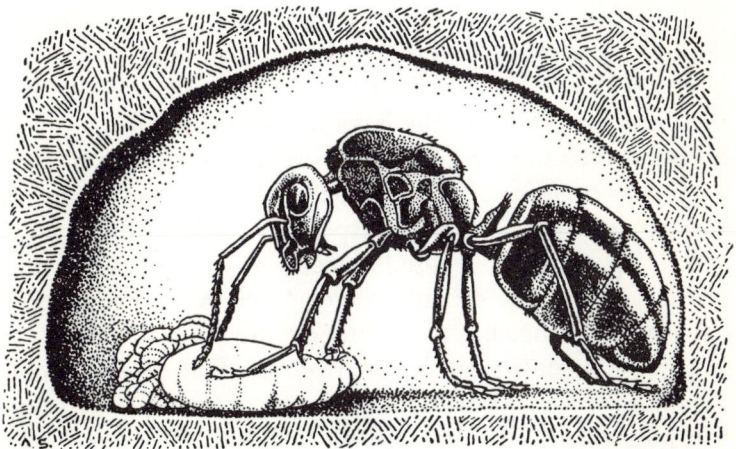

Fig. 3. A queen ant starting a new colony. (From T. C. Schneirla)

and as the eggs hatch she begins feeding the larvae with her own saliva. The larvae pupate and emerge as very small workers and these will now go out and get food to feed the queen and will feed and take care of succeeding broods. The queen may appear to lose her instinct as soon as she has help, and from then on she concentrates on the making and the laying of eggs.

The coordination of a colony of social insects is achieved in a number of ways. We have already mentioned the exchange of food materials which act as a bond between individuals, but in addition this touching and palping of one another often imparts information, that is chemical information. Communication of all sorts is now best known in the bees, so let us take our examples there. I will not devote a long section to the bees, however, because the famous discoverer of their language, Karl von Frisch, has already covered the subject in his admirable little book. Here I will merely point out a few of the salient facts.

As has already been mentioned, a bee returning from a flower bed tells the other bees about the kind of flower by chemical signals—they smell the scent that clings to his body. He will allow himself to be poked and stroked so that all those that are about to go out and hunt nectar and pollen will know the smell of a good source of food. But even more remarkable, the incoming bee can tell the others both the direction and the distance. The direction is given by the orientation of a dance which the bee performs, for the dance is oriented with respect to the direction of the sun. The sun may be used directly as a compass, or any portion of the blue sky may serve as well, for the light from the sky is polarized, a phenomenon to which the bee's eye is especially sensitive. And so from a small patch of blue which he sees from inside the hive he can accurately gauge the position of the sun and make his dance accordingly. The distance is indicated in two different ways: if the source of food is close by the incoming bee does a special whirling dance; if the distance is greater than two hundred yards the rate of the orientation dance is an index of the distance. By knowing these things von Frisch was able to predict the exact direction and distance of a particular food supply with the same accuracy as the bees. This is certainly communication of a high order of complexity, far greater than had been previously suspected in insects.

There are even instances of auditory communication between adults and their young, seen especially clearly in certain social wasps. Here the larvae in the cells have a salivary fluid that the adults are eager for and this they take essentially in exchange for the pellet of food they give the larvae. The wing vibrations of the adult wasp serve as a stimulus for the larvae which all protrude from their cells and at the same time commence salivating much like Pavlov's famous dogs. That the wing beat is the stimulus can be shown by giving a shrill whistle (or a similar vibration) and this also will evoke the response. One is certainly reminded of a group of baby robins in the nest, all opening their gullets in response to the presence of the mother, but the baby robins do not offer anything material in return and also they apparently respond to different stimuli.

Lastly I should like to mention the nests of social insects, for these, like the structures of the beaver, become a central part of their life history and activities. The nests vary from small groups of cells to huge and complex structures. We are all familiar with the combs of a beehive and the large acorn-shaped paper nests of wasps. Termites not only live in the hollows of wood that they have carved out, but also build, as for instance in the Belgian Congo, great tall and irregular stalagmites, twice the height of a man. Many ants burrow in the ground and tunnel over large areas, while some make their nests in trees, in hollow branches or in balls of leaves that they cement together. The tailor ants do this in teams of workmen, some pulling the leaves together and others squeezing their larvae like tubes of paste, to provide a sticky cement which glues the leaves (Fig. 4). The variety of architecture is great and often ingenious, and in each instance serves to aid them in their feeding, their reproduction, and to help protect them from their enemies. Their nests are fortresses with many advantages. There are,

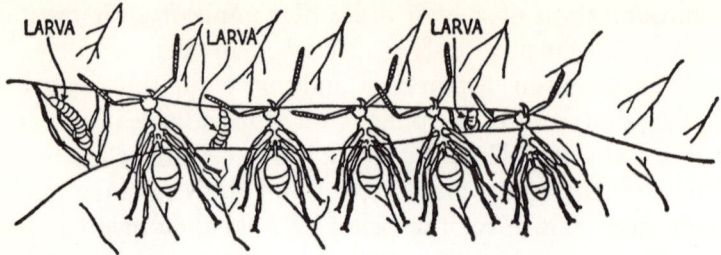

FIG. 4. Tailor ants making a nest by pulling leaves together and sticking them with silk spun by their larvae. (From O. W. Richards, after Doflein)

however, some ants, namely the army ants, which do not stop long enough in any one spot to build a nest, and let us now examine them in more detail.

ARMY ANTS

THE army ants, which are also called legionary or driver ants, are found in many warmer regions of the world. These ants are strict carnivores and their habits and life history are centered about their flesh eating. They attack in great hordes and pillage the country-side seeking other insects and even small mammals if they are lucky enough to catch them, and they savagely tear their prey to bits and carry off the meat in small pieces as they stream on in search of new victims. As Wheeler says, they are "the Huns and the Tartars of the insect world."

When they visit a house in the tropics they are considered desirable because they clean the house of vermin, although should this happen to me my feelings on the subject are likely to be mixed. Some species attack by night and their approach is heralded by a scuttling of the cockroaches, mice, lizards, and rats in the house, a commotion that cannot fail to wake even a sound sleeper. It is then wisest to grab one's bedding and leave the house (being careful where one treads) for the army ants are now in full possession and one can only wait until they have cleaned the house. Some people place the feet of their bed in pans of vinegar water so that the bed may be a safe island refuge, but then the ants have been known to chase some household insect onto the ceiling right above the bed and drop in a mass with the victim, right on the unhappy man who thought himself so safe in his four puddles of vinegar-water.

The army ants of central America make their hunting forays in the daytime. The light itself is of no value for the purpose of seeing objects, for the eyes of these ants (with the exception of the fertile male and female), are so rudimentary and atrophied that they can do no more than distinguish the difference between light and dark. They spend the night in a bivouac, which is a pendulant cluster of ants like a swarm of bees, slung from some low fallen branch, and when the light of early morning comes this mass of 100,000 to 150,000 ants starts to shimmer with activity as they descend to the ground and begin walking about after one another in whirlpools and eddies. Then suddenly the simmering pot boils over and a column of ants, an inch wide and six to a dozen ants abreast, pushes out into the forest, with all the rest following in a long file (Plate VI). The sterile female workers are of four sizes or castes in these ants, the large soldiers down to the small minim workers (Fig. 5). The smaller castes are successful in hug-

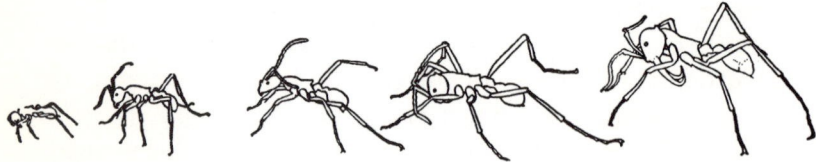

Fig. 5. A series of different-sized workers of army ants. (From T. C. Schneirla)

ging the center of the column, and the larger ones crowd in as much as possible, but find the footing more stable on the outside edge, and so the column itself, by this happenstance, appears to have a military order. Some of the ants may be seen carrying larvae (which they hold in their jaws with the larva slung beneath them between their legs) for they must bring the whole household. Most important of all, somewhere near the tail end of the column, the queen will be seen marching, surrounded by a group of especially attentive and excited workers, showing due respect for their ruler and their mother.

The front of the column now spreads out and becomes the attacking group. It may assume various shapes, an eliptical knob, or it may spread out like a fan, and in either case it will be constantly pushed forward by the pressure from the marching column behind. The forward movement of the front edge is often not smooth but irregular and rocking from side to side and this motion serves to out-flank any prey that is attempting an escape. When they do trap some terrified grasshopper, those in the vicinity come and tear at it with their sickle-shaped jaws and soon nothing is left but small fragments carried by many ants as they hurry on. The advancing swarm may be as broad as fifteen yards and before it the victims can be seen trying to escape. Occasionally small birds may be hovering about to eat in-sects that the ants have flushed. Sometimes they will an-nihilate the nest of another species of ant, or the nest of a wasp; sometimes they will find a pile of leaves and humus and clean it of grubs. As the day moves on they again col-lect in an orderly column and by nightfall they have found a new site to bivouac, perhaps some three hundred yards from the place where they spent the previous night.

It was known that there were periods when the ants would move and camp at a different place each night, and other periods when they would find some hollow in a tree and come there for many nights in succession, but the reasons for this have not been understood until recently. T. C. Schneirla, an animal psychologist from the American Museum of Natural History, has spent many years making detailed and careful observations of the habits of army ants, most especially those of Barro Colorado Island where Car-penter studied the howling monkeys, and Schneirla has written some excellent and meticulous studies on their social habits.

First of all he found that the nomadic periods involving the daily changing of the bivouac and the relatively sta-tionary periods in which some tree hollow is used every night as a camp site alternate in a very regular fashion. The species (*Eciton hamatum*) on which he made the majority of his observations wanders for a period of seventeen days and then remains in one place for nineteen to twenty days, and this is true all the year round, in both wet and dry seasons. Thus no external influence of weather or moon periods seems to cause this cycle, but rather the cause is to be found, as Schneirla showed, in the reproductive cycle. The stationary bivouac is the place where the large and mature larvae, which have been carried about on the previous raids and fattened on the spoils of war, make their cocoons and become relatively dormant as they are becoming transformed inside their silken capsules into mature ants. The onset of cocoon-making coincides with the onset of the sedentary phase, and not only do they keep the same bivouac every evening but the raids themselves are smaller and less effectual. Another event occurs during this period: the queen's abdomen becomes greatly swollen and after a week the fact that she is pregnant is no longer possible to conceal, for she has swollen to five times her previous size (Plate VI). Then for an interval of five to seven days she lays eggs, twenty to thirty thousand of them. The workers help to deliver the eggs and feed the queen, always rushing about her in a somewhat frenzied fashion. Soon the eggs hatch and the minute larvae are also attended to by the nurse workers; by the time the twenty days are up the larvae are large and hardy enough to be transported from camp to camp as they enter the nomadic phase.

The stimulus to travel again is apparently produced partly by the wriggling of the larvae which excites the workers, but also at the end of the stationary phase the pupae begin to hatch (helped by the excited handling of the workers) and the new callow workers emerge. This increases the frenzied activity of the colony to an even higher pitch; they can no longer contain themselves and they stream off to begin their wandering and hunting.

The broods of which we have been speaking are all worker broods, but there is the question of the production of new colonies, a matter which involves the formation of new males and new queens. Periodically, and this phase does appear to be affected by the weather, the queen lays eggs that develop into males and females, usually sometime early in the dry season. The males that emerge may be quite numerous, two to three hundred in number, but there are only about half a dozen females, the presumptive queens. It may be that there were more larval queens and that they did not survive, for there is competition even among the remaining handful and only one will live to start a new colony.

When the new queens emerge from their cocoons there is, according to Schneirla, a great commotion among the workers, for their loyalties are actually split. Each queen, including the queen mother, will be surrounded by workers, some showing kind attention and others apparently irritated and petulant with her. This over-attention will be the death of the weaker ones, and finally only two are left, usually the old queen and one new queen. Then they separate and their separation produces a great split in the workers, each following their preferred ruler. The colony actually divides in two and becomes two separate colonies. It is suggested by Schneirla that this recognition of the various queens starts when the new queens are still larvae and reaches its climax only when they emerge from the cocoons.

No particular attention is paid to the males. These large, winged creatures wander off and are accepted in new colonies of virgin queens so that a certain amount of cross-fertilization between colonies is achieved. Only one male, of course, ever reaches a virgin queen, and he fertilizes her. For him, this is the end and he soon is ignored and overshadowed by the great matriarchy.

A particularly interesting problem arises if through some accident a queen is lost and the colony is without its means of perpetuation. Apparently then the ants are willing to devote their attention to the queen of any other colony, and they are accepted by the workers. So a queenless group will lose no time in fusing with a normal colony.

It has by now, I think, become amply obvious that the cohesiveness of the group depends to a very great extent on the coordination among the individuals. It involves, for one, recognition of particular queens. This is achieved through the chemical senses, for in these essentially blind ants smell and taste discrimination are certainly very acute. Curiously enough there is a difference in the smell of a mass of workers, and the smell of a queen to the relatively insensitive nose of man. The workers have a smell "reminiscent of potato blossoms" (although often it is overshadowed by the fetid odor of the food), while the queen has a more "delicate, indefinable odor." Of course the ants can further distinguish between particular queens, as their reaction to the new queen certainly indicates. Therefore recognition of one individual, the queen, seems to serve as a strong binding mechanism, and should the queen die they are quick to focus this attention on another queen. She is not just a symbol of the center of their organization, but because of the response of the workers to her odor she physically coordinates the colony into a unit.

The chemical sense has a further effect on the coordination of their marching and foray movements. The workers are endowed with a strong instinct or response to follow the track, the scent trail, of their fellow workers. There are no scouts or leaders at the head of a column, but merely a mass of ants intent on following one another. This means, when more than one hundred thousand ants are involved, that the pressure of followers will be great enough to literally push forward the irresolute front rows. In fact the leaders will be continually replaced by new ones as they surge forward from behind and as soon as the new leaders falter, new ones again surge past them. Both Wheeler and Schneirla have shown that this slavish instinct to follow can in the laboratory, or occasionally by some freak in nature, result in their destruction. Should they begin to follow one another in a circle, around a dish of water for example, they will always be in the ideal position of following a fresh scent, and never be pushed out front. But they cannot march forever and they will eventually die in their tracks. There is no power to reason; they have been trapped in their own simple set of stimuli and responses.

Intercommunication is also illustrated by their exchange of food. There is to some extent a division of labor among the different castes of workers. The large ones are the most active in foraging while the smallest ones are the most active in the care of the queen and the young. The queen herself does no attacking, and therefore the queen, the larvae, and to some extent the smaller workers must live on the food that is shared by the larger workers. The feeding involves a distribution, a consideration of the colony as a whole. From the point of view of natural selection this is an especially obvious and necessary mechanism for if the labor of food-getting is to be divided, then there must be sharing or else only the sterile workers will survive and be incapable of perpetuating themselves. This is admittedly rather an obvious point but it illustrates clearly the fact that the greater the division of labor of a living unit, the greater will be its cohesiveness.

Among ants army ants are peculiar in not having a nest. It is true that during their stationary phase they live in a hollow of some sort, but this is of little consequence, for the fact is that no matter where they camp, each night they build their own nest and this nest is made up of the worker ants themselves (Plate VII). There are layers and layers of workers which hang one from another, and within the center of this ant-made nest lies the queen with her brood and small nurses. There may even be shimmering, temporary corridors which facilitate movement inside and help the aeration of the mass. But the whole structure, and it does have shape and form as well as differentiation of layers, is made of many thousands of ants. The nest in this case is a vivid demonstration of cohesiveness and mutual affinity.

TERMITES

THE termites are members of a totally different group of insects from the bees, ants, and wasps; they are more closely related to the cockroaches. This fact is important both in understanding the differences between their societies and those that we have already discussed, as well as emphasizing the point that societies have arisen independently many times during the course of evolution. The most noteworthy difference found in the termites is that they do not follow the sequence of egg, larva, pupa, and adult; instead they have a so-called incomplete metamorphosis with no true maggot-like larva, but a succession of nymphs (diminutive forms only slightly modified from the adults in appearance) which go through a series of molts as they gradually change leading finally to the last molt when they become adults (Fig. 6). This means that except when nymphs are extremely young they are not helpless like larvae, but may serve as workers. Another great peculiarity of termites is the fact that they are not composed merely of females, but the sterile workers are of both sexes and the queen is always accompanied by her king.

Termites are primarily tropical, although there are many species in various parts of the United States. As is well known, they have a particular propensity for eating wood. Buildings, books, telephone poles—anything made of cellulose not treated with some resistant chemical will be voraciously consumed. They are especially dangerous because they have an aversion to being out in the open—either the light or more probably the moisture conditions of the outside are unpleasant to them—so they carve out the insides of beams and panels, leaving a thin veneer of untouched wood. A strong and healthy-looking house may suddenly collapse when the massive wooden supports have been slowly hollowed out and have no more strength than an empty cardboard box. The desire to keep covered and shielded from the fresh air is manifested in many ways. If

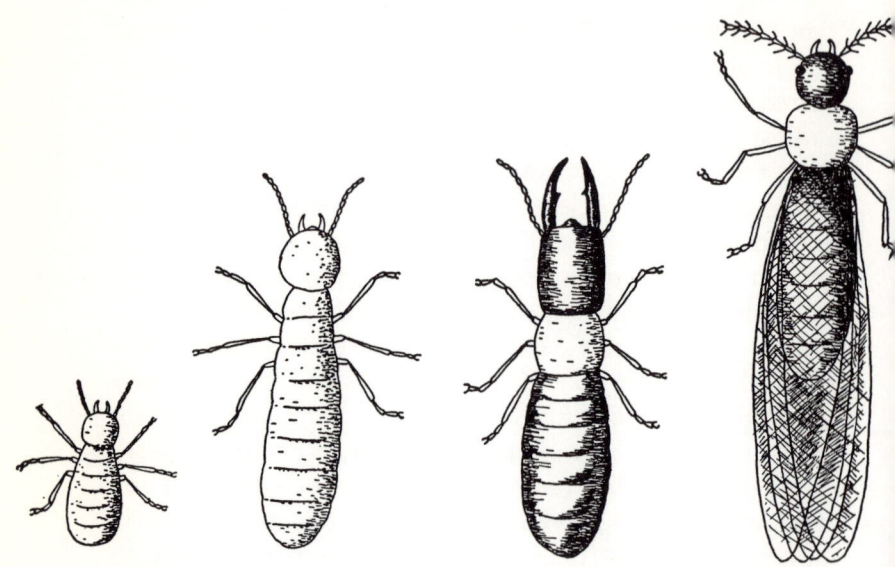

Fig. 6. Some different stages in the development of one species of termite. From left to right: two nymphs of different ages, a soldier, and a queen.

a house stands upon a high concrete basement, the termites will make small tunnels with saliva and wood paste from the ground, up the concrete walls to the wood of the house; this is a covered bridge which brings them across unfriendly and inedible areas. It should be pointed out, and this is especially evident in the tropics, that termites do not exist solely on the works of man; the fact that any tropical jungle will be clean and bare of fallen logs can be ascribed directly to the activity of termites.

Cellulose is no easy substance to digest. It is made up,

it is true, of sugar molecules hooked up end to end, but the strength of the bonds is so great that the digestive enzymes of human beings, for instance, are quite incapable of splitting the molecules and obtaining the energy of the sugar. Certain microorganisms, protozoa, and bacteria have special digestive juices that can perform this rugged feat; and cows or rabbits, which depend to some extent on cellulose, have a special vat in their intestine where bacteria can split the cellulose and allow the animals to use the sugar. The bacteria living in these animals are friends in that they are essential for digestion, and the termites have a special menagerie of complex protozoa in their gut which allow them to get nourishment from the wood. Each time the nymphs molt they lose that part of their gut containing the protozoa, and they must get some new protozoa from another termite, which they do readily by licking one another. Even the king and queen when they begin a new colony must bring their domesticated protozoans with them or else the colony will never start. This process has been going on for so many years that the species of protozoa are characteristic for a particular species of termite.

There are many different kinds of termites and a great variation in the kinds of societies they produce, but invariably the colony is centered around the king as well as the queen. For the nuptial flight the winged presumptive monarchs (or more correctly the presumptive parents of a very large family) will fly together when certain very specific atmospheric conditions are attained, and finally descend to the ground, quickly shedding their wings which snap off at their breaking points. The female takes the lead with the male following close behind and together they dig a burrow in which they close themselves off from the outside. In this bridal suite they will mate and after some time a few eggs will be produced. The starting process often takes a long time and the building of a large colony may be a matter of years.

In some tropical forms the abdomen of the female becomes enormous and she lies, like a giant grub unable to move, in a disc-shaped royal chamber, with the relatively minute king ever attentive by her side. Around her there will also be many workers of different sizes and even some soldiers ambling about ever ready to protect her in the advent of danger. The nurses feed her and take the eggs from her as well as lick and clean her constantly. The queen is really but one huge egg factory; once the colony is well established she may produce over ten thousand eggs in a day.

As long as the king and queen function, all the workers and soldiers remain sterile. But should either or both of the royal pair die, then they will quickly be replaced by secondary reproductives, workers who upon molting have a slightly altered, darker appearance and are capable of manufacturing fertile sperm and eggs. Sometimes there are more than one pair of secondary reproductives to replace the loss of the original parents. In some extremely large colonies which may involve hundreds of thousands of individuals an interesting situation occurs where the old king and queen are the sole fertile forms for the whole central part of the nest, but in numerous regions at the outer edge of the colony the dominance of the parents will not be felt and secondary kings and queens will arise. There is apparently something in the presence of reproductives which prevents nymphs from molting into reproductives.

The problem of the balance of castes, the balance of the division of labor within a colony, is one that has fascinated biologists for many years. It is now generally agreed that the fertilized eggs are genetically the same (except that some are potential males and some potential females) and the differences between the castes are somehow produced either by differential feeding of the nymphs or possibly by specific inhibitory or stimulatory substances that are passed from one individual to another. As in other social insects there is a constant exchange of saliva between termites, and they also lick one another's bodies as well as the anus of their colony mates; so there are unlimited possibilities of passing specific substances, or "social hormones" as they have been called, from one individual to another. One might postulate, for instance, that the king and queen of the large colony previously mentioned give off a substance which inhibits any nymph from becoming a fertile secondary reproductive. But if the colony reaches too great a size, the single royal pair do not give off enough of this inhibitory hormone to extend to the outer edges of the vast expanding colony, and in those peripheral regions secondary reproductives do arise.

The actual experimental evidence for such a situation is unfortunately meager and inconclusive. There are, however, some most interesting recent experiments of a Swiss zoologist, Martin Lüscher, which show that there are regulatory mechanisms not only for the production of secondary reproductives but for their maintenance, once they are produced. He divided a small colony of forty termites into two cages adjacent to one another, but separated by a screen which allowed the termites to touch their antennae. In one of the groups of twenty there were a king and queen, but in the other there was neither. After a time, however, a new king and queen did appear in the sterile side showing that whatever prevents the production of fertile forms cannot be transmitted by touching antennae. However, the workers were quite dissatisfied with the new royalty for they somehow sensed the presence of the older royalty through the screen (presumably by antennae touching) and proceeded to kill and eat the new pair. Of the eighteen remaining termites two others later developed into fertile forms and were again eaten, and the process was continued until there were very few termites left in the cage. To show that the antenna-touching was necessary, Lüscher placed a double screen between the two groups, preventing the antennae tips from reaching across, and then the new royalty was not attacked and both colonies lived in stability, each with its own king and queen, for a long period of time. This shows that there is a separate mechanism for the production of royalty and for their maintenance, and furthermore that the maintenance is dependent upon communication between antennae. It is not clear how the signal is passed from one antenna to another; it is quite reasonable to assume that some substance is passed on these chemical-sensitive appendages.

Aside from the reproductives there are a number of castes among the workers and soldiers. In some species of termites there are two classes of workers and two classes of soldiers. The soldiers often possess, as did the soldier ants, large strong mandibles, but in numerous species there are soldiers whose head region has become so modified that they have a snout shaped like a small syringe. These so-called "nasute" soldiers produce a paralyzing liquid which they can eject from their syringe-like beak, making an effective weapon of war (Fig. 7).

The nests of termites, or termitaria, vary greatly in size and complexity, but the most interesting ones are the large tropical mounds (Plate VIII). In the center of these termite cities the royal chamber with the swollen queen may be found, and radiating from this there is a vast network of chambers and tunnels, all joining in a complex and irregular fashion. The building material of the termitarium is processed wood and dirt, cemented with saliva and excreta to make a good rigid wall. They rely upon the strength of the nest to keep off their enemies, and within those dark, humid passageways, sealed off from the unfriendly world, they carry on their complex and integrated social life. In the struggle for existence they have isolated themselves in dark fortresses; they have sealed themselves off in an environment in which they are the rulers and in which the competition is negligible.

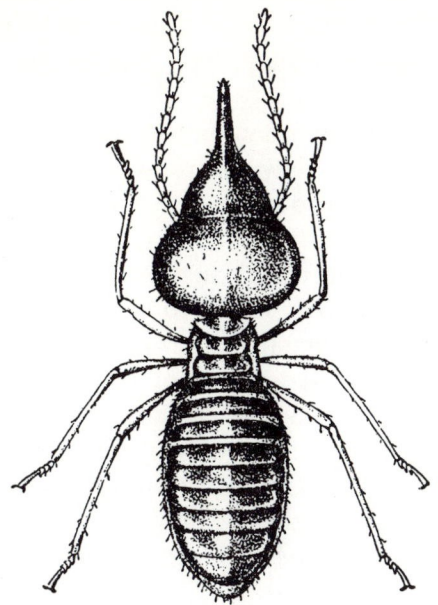

Fig. 7. A termite nasute soldier. (From N. Banks)

COLONIAL HYDROIDS

On a warm sunny afternoon, in the middle of the summer, I can remember lying down on the hot boards of a floating dock, and by peering over the edge and shading my eyes with my cap, I could see the cool green sea water, and along the side of the wharf there seemed to be many different kinds of seaweed, some red and some brown, undulating with the movement of the water. Among this weed there were small tree-like clusters an inch or two long, many little thin white branches often flecked with spots of red. Each little tree is a whole colony of small animals that are connected one to another; they are colonies of hydroids. But before discussing these colonial forms it is necessary to learn something of the basic structure of hydroids in general, and for this the fresh-water polyp, hydra, serves as an excellent example.

Hydra was first discovered by Antony van Leeuwenhoek in the latter part of the seventeenth century, but it was poorly understood and even thought to be a plant until the gifted zoologist from Geneva, Abraham Trembley, in 1744 published his careful study of that organism. Trembley was working as a tutor in the grand house of his rich patron, Count Bentinck in Holland, and in the ditches and canals about the estate he found what he called a "polyp with arms in the shape of horns." With painstaking precision he described the movement, the feeding, the reproduction of this animal that, save for the microscopic detail, remains true and unaltered to this day. He had no effective microscope and therefore could not see its minutest aspects, but he had the rare gift of describing solely what he observed. As we shall see shortly, this polyp gives rise to young by budding, and Trembley noticed that these buds may be quite numerous before separating from the mother, giving a many-headed appearance, like the hydra of the Greek myths. To make the name even more appropriate, after discovering that this polyp will regenerate a new head, he carefully made a number of incisions in the cut stump and, to use his words, "after a few days I saw in it a prodigy scarcely inferior to the fabulous Hydre of Lernaea."

The shape of hydra, which stands an eighth to a half-inch high, is roughly that of a delicate vase in which the base is narrow, the upper portion slightly rounded, narrowing again at the upper lip which is festooned with tentacles, a fringe of delicate, muscular, serpentine threads (Fig. 8). Like a vase, hydra has only one opening to the internal gut cavity; the food captured and paralyzed by the tentacles with their stinging cells is pushed down through the opening, and once all the nutrition has been washed out of the prey by the digestive juices, the remaining useless excreta is ejected back out the same opening. Even though this opening has a dual function, nevertheless, for propriety we call it a mouth.

The body wall of hydra is a simple structure. It consists primarily of two layers of cells: the outer and the inner (Fig. 9). There is also a thin middle layer which lies between, but this layer is not a neat row of cells as are the other two, but a thin layer of adhesive material in which there may be a few cells or parts of cells. The inner layer possesses many gland cells which secrete the digestive juices. The food itself, once in solution inside the gut, is directly absorbed by the cells of both layers, for the body wall is so thin that the food can easily diffuse to all parts.

The power of muscular contraction in hydra is striking, for if one irritates an extended individual with a needle, it will quite rapidly contract into a minute ball; even the tentacles, which are very fine and thread-like when extended, look like short stubby pegs when contracted. This con-

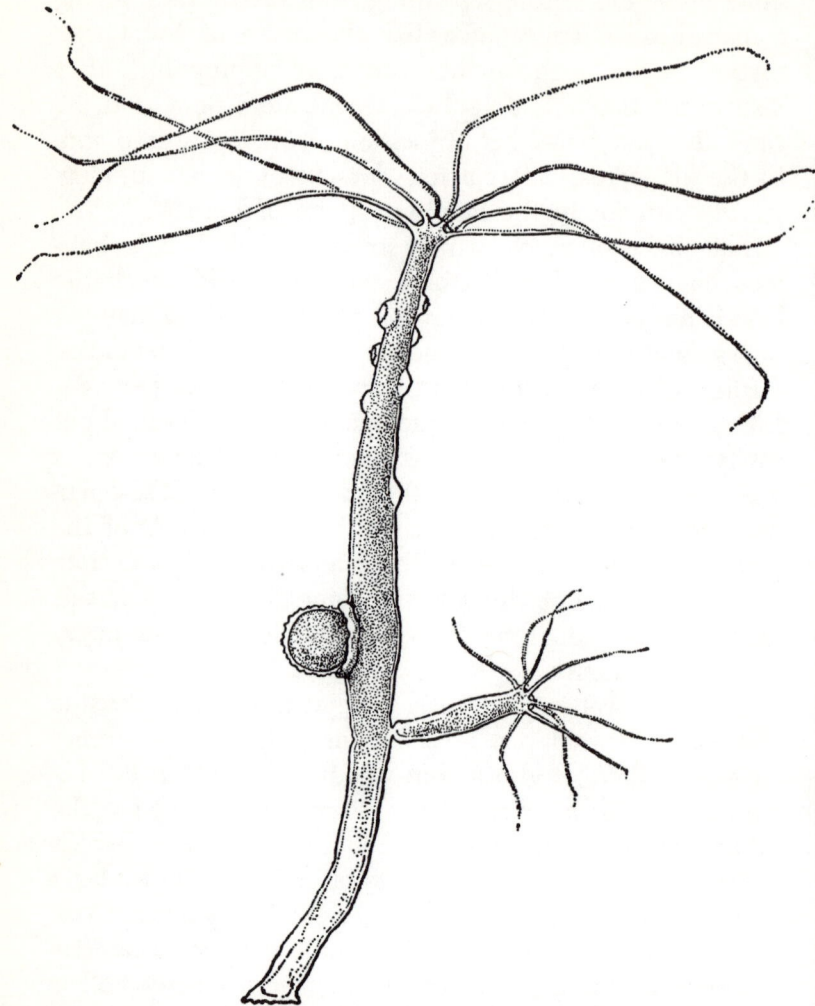

FIG. 8. The common green hydra showing an asexual bud on the right-hand side, testicles on the upper region, and a partially developed egg emerging as a large sphere on the left-hand side. (From P. Brien and M. Reniers-Decoen)

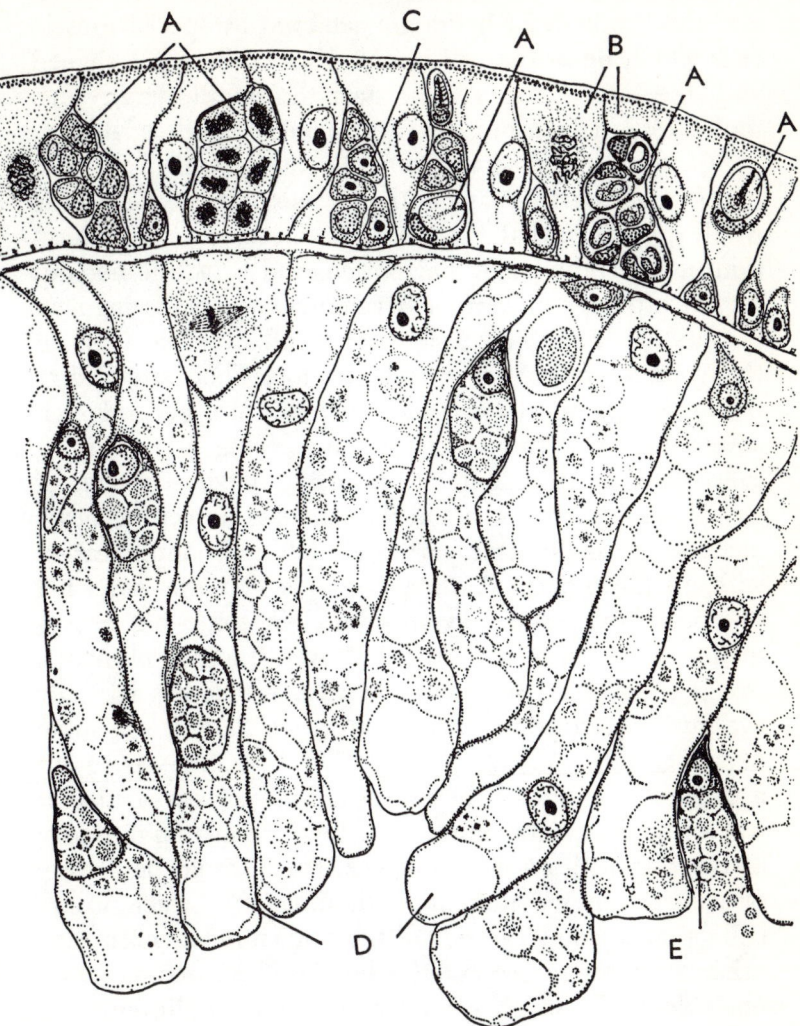

Fig. 9. A cross-section of hydra showing the three layers (the outer layer is above). A shows the progressive stages in the manufacturing of stinging cells; B, the contractile covering cells; C, small interstitial cells; D, the large inner cells; E, the gland cells, some of which are liberating digestive juices into the gut. (From P. Brien and M. Reniers-Decoen)

traction and in fact all the other movements of the tentacles and the body of hydra are achieved by special muscle cells which lie in the cell layers. These cells are T-shaped with the roof of the T lying near the middle layer. This roof is essentially a muscle; it can contract and expand.

But it is most obvious from the way in which the contraction takes place, namely that if one part is touched the whole contracts, that there is some means of communication between the parts. This is achieved by a true nervous system, although an extremely rudimentary one. Lying near the middle layer there are scattered all over the body of the hydra a series of nerve cells with long fibers attached to them, and these cells and their fibers form a network that covers the whole surface of the animal (Fig. 10). This so-called "nerve-net" is different from the nervous systems of any of the higher animals in that there is no central headquarters, no brain. One part is directly in communication with all the others. The manner in which this nerve-net functions was beautifully illustrated in some old experiments of the zoologist Alfred Mayer. Instead of hydra, which is so small, he used a jellyfish which has basically the same structure as hydra, and as we shall see in more detail shortly, is directly related. The first step was to cut a circular hole out of the center of the jellyfish so that it was a ring, like a doughnut. If he then stimulated one edge of the ring with an electric shock, there would be a wave of contraction radiating in both directions which would collide on the opposite side of the ring and extinguish each other. In order to prevent this he placed a piece of ice on one side of the starting point so that one of the impulses disappeared by entering a numb area, and then the ice was removed so that the area would again become normal. The wave of contraction going around the other way now had nothing to oppose it and it continued to go around and around the doughnut for many hours, not stopping until the mutilated jellyfish died. The nerve-net does not integrate or sort out impulses as does a brain, it just indiscriminately passes on information.

Sexual reproduction in the fresh-water hydra is extremely atypical and degenerate, and therefore I will not stress it;

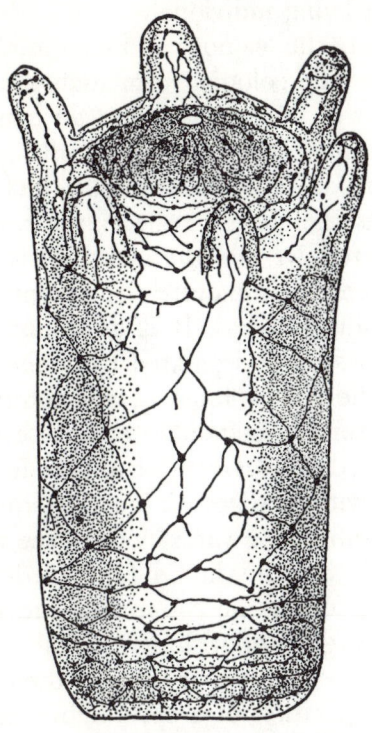

Fig. 10. A diagram of the nerve net of hydra. (From L. H. Hyman after Hadzi)

but the asexual budding, a most usual mode of propagation, does have some basic features in common with the colonial forms. There is a region on the lower part of the abdominal swelling of the vase-like body where a bump may begin to appear. This protrusion or blister soon develops an anterior corolla of tentacles and appears eventually to be a small edition of the parent from whose flank it has burgeoned.

Its inside gut cavity is in direct continuity with that of the adult so that in a sense these two form a colony. In the colonial forms this is exactly the way the colonies arise, by repetitive asexual budding; but in hydra itself, which is degenerate and anomalous, the bud separates and wanders off to be a free-living individual.

Let us return to the sea now, and remembering the hydra, let us look at a true colonial form such as Obelia. It consists of a group of hydras or polyps all connected to one another like branches of a tree. Not only are the outside, middle, and inside layers continuous, but so is the gut cavity, which is therefore a tube connecting all the individuals. It also follows that the nerve-net is continuous; in fact the whole colony is in every respect one flesh, made up of many individual hydras. It is as though in hydra the daughter colonies never separated from the parent.

Feeding in the colony must be a communal affair. For if one polyp is fortunate enough to capture some food, say a minute shrimp, then it will share this food with all its neighbors. For with its digestive juices it liquefies the nutriment in the shrimp, afterwards ejecting the useless remains out the mouth, and this liquid food feeds not only this polyp but passes down the communal gut, the communal pipe system, and gives nourishment to all the other polyps as well as the living stems that connect them. In a way this is a mutual cooperation of a very high order; it does not involve senses and instincts but it is the direct result of the construction of the colony. The bonds between the individuals can no longer be across a space for they lack a nervous system sufficiently organized to cooperate at a distance; they cannot be aware of one another's presence, they cannot smell, see, and feel each other and instinctively act harmoniously. Separate fresh water hydras show no social tendencies whatsoever. The only way in which cooperation can be achieved here is by being physically tied together so intimately that they cannot escape being cooperative. No doubt this has advantages of survival, and by this mutual aid they protect one another, but they can only achieve this advantage by being corporally bound.

The reproduction of Obelia involves an alternating sexual phase and an asexual budding or growth phase (Fig. 11).

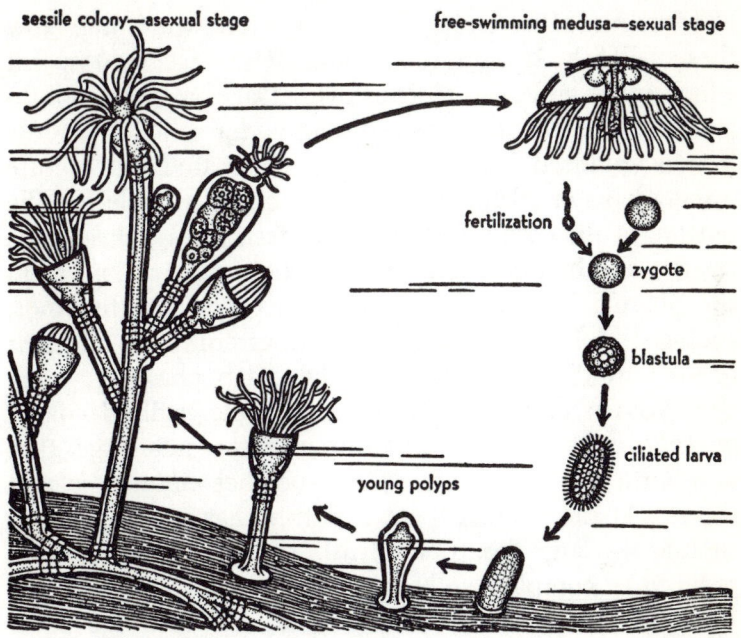

FIG. 11. The life cycle of Obelia. (From R. Buchsbaum)

In the latter, just as we saw in hydra, a bud develops into a new polyp, but in this case, as I have already said, the off-spring does not separate from its parent. So the reproduction of the individual members that make up the colony is by asexual budding and there is no question here of sexual attraction as a force responsible for their social existence. There are among the many species of colonial hydroids a great many ways in which buds arise, resulting in a great variety in the overall shapes of the colonies and types of branching, but in every case the principle and the result is basically the same, the propagation of new individuals by vegetative buds.

In Obelia there are certain specialized polyps that appear when the colony is mature. Instead of being simple-feeding hydras, these have a series of projections or buds which give rise to small, delicate, beautifully transparent little medusae that escape and become free-swimming. Now these jellyfish in turn soon manufacture egg and sperm which unite and after a series of cleavages a small mouthless and tentacleless free-swimming hydra called a planula larva is produced. This planula, which is coated with minute waving hairs or cilia, swims about and finally settles on the bottom of the ocean floor, attaches itself firmly, and begins to grow and bud off polyps. So the whole colony stems from the planula larva and, in fact, students of the hydroids, zoologists who have studied them carefully in all their forms and stages, believe that the polyp phase of which the hydra is one example and the whole hydroid colony another, is in reality an infantile or embryonic stage, that it is but a modified planula larva and that maturity is only reached at the medusa stage. It is true that in many forms, such as the large group of jellyfish, the polyp stage is often reduced or completely absent. However, in many hydroids the medusa stage is reduced so that the medusae never leave the colony, in fact they hardly even resemble medusae but are merely sperm or egg sacs attached directly to the branching colony. So we see in the life cycles of these forms that in some cases certain stages are more important than others, and specifically in the colonial forms it is the embryonic polyp phase that has taken on such a variety of shapes and forms and such a complex communal existence.

In the polyp colonies we already have an example of division of labor, for the majority of the polyps feed and this is their sole function, while a certain few, in a mature colony of Obelia, cannot feed but have the sole function of reproducing and producing medusae. They obtain their energy, their food, through the gut canal which connects them to the feeding polyps, and in return they reproduce new colonies which the feeding polyps are incapable of doing.

In other forms, as for instance Hydractinia, the division of labor is even greater. This hydroid does not stand erect but covers the shells of snails that are inhabited by hermit crabs (Fig. 12). It forms a close mat over the surface of the shell, a mass of tubes connecting the various polyps,

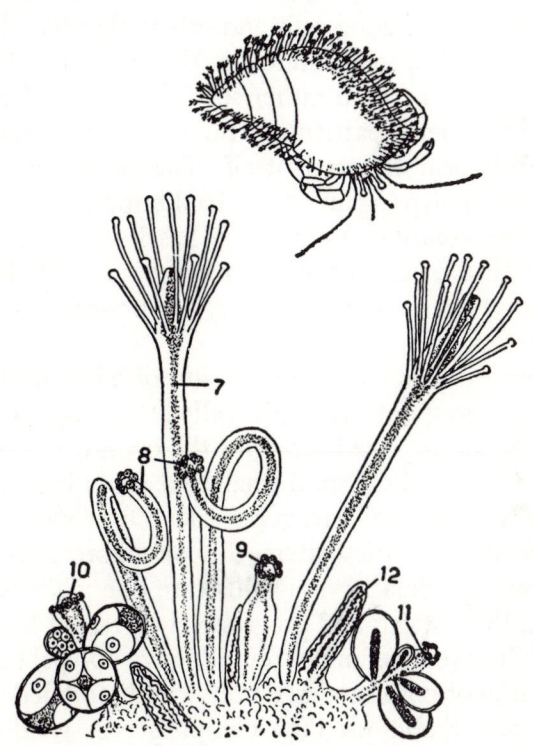

Fig. 12. Hydractinia. *Above*: Its appearance on a snail shell inhabited by a hermit crab. *Below*: A cluster of polyps showing the different types; 7, a feeding polyp; 8, protective polyp; 9, 10, 11, immature, male, and female reproductive polyps, respectively; 12, spines. (From L. H. Hyman)

which stand erect like small white frosty bristles. Among these there are many feeding polyps, each closely resembling hydra and each ready to catch some morsel that the clumsy hermit crab might let escape from his mouth. Besides these there are the reproductive polyps which bear the medusae, but since the medusae do not escape in this form these polyps will either bear the opaque white sperm or fat greenish eggs directly, depending on their sex. There is another kind of polyp, protective polyps, and these exist in two forms: there are some with mace-like knobs at the tip packed with stinging cells and others that look like long thin fingers. Of special interest is the fact that these protective polyps take up a rather special place in the colony, near the edge, being particularly numerous about the mouth of the shell near the crab itself. The protective and the reproductive polyps again do not feed but receive their food through the communal gut. Each has its specific function and as with Obelia each works for the good of the whole community, but only because they are bound together and cannot escape.

The most extreme cases of division of labor are found in a peculiar group of hydroids called the Siphonophores. These animals are free-floating in the open ocean, many of them relatively delicate and small, but some large and conspicuous, such as the Portuguese man-o'-war which we find in our South Atlantic waters. Siphonophore colonies also arise from a planula larva, in this case one which never attaches to the bottom and progressively gives off buds, some of which become feeding or reproductive polyps, but others form a series of both modified polyps and modified medusae that are remarkable for their complexity. A good example is Agalma, a form which has the advantage for us of being somewhat strung out so that it is possible to see its various parts (Fig. 13). At the apex there is a small float which is a modified attached medusa that confines a certain amount of gas and serves to keep the colony afloat. While the float

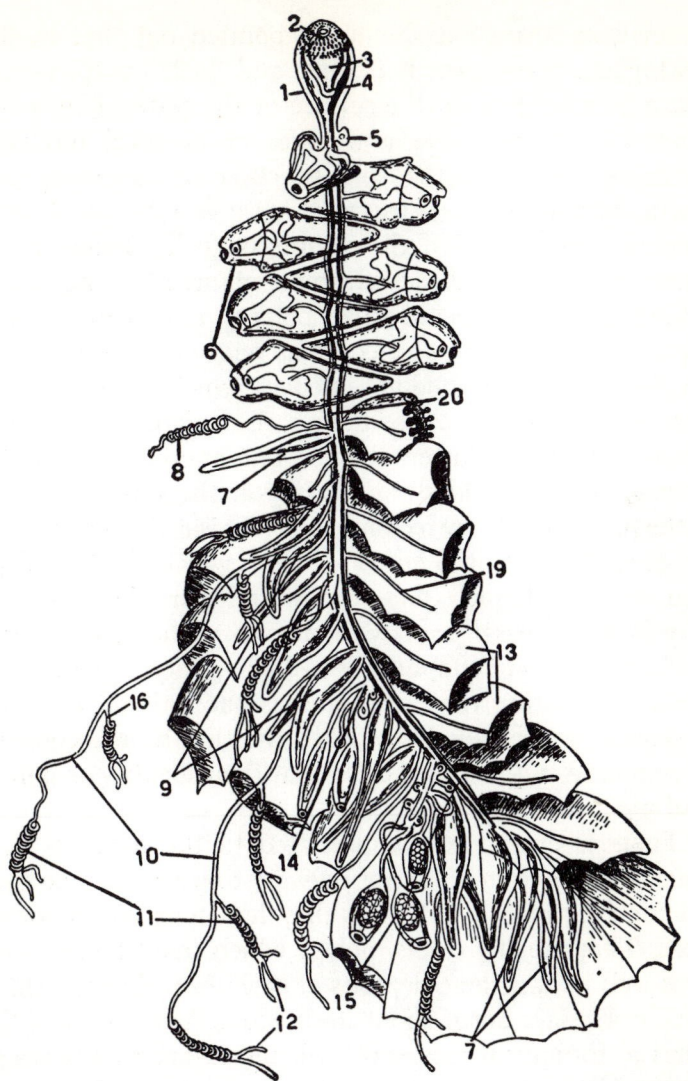

FIG. 13. The siphonophore Agalma. 1, the float; 2, pigment in the float; 3, air sac; 4, funnel; 5, the budding zone for the swimming bells; 6, swimming bells; 7, the "feeler" polyps; 8, their tentacles; 9, the feeding polyps; 10, their tentacles; 11, area of stinging cells; 12, end filaments; 13, protective modified medusae; 14, sperm-bearing male polyps; 15, egg-bearing female polyps; 16, lateral con-tractile branches of the tentacles. (From L. H. Hyman after Mayer)

is small in Agalma it should be pointed out that in the Portuguese man-o'-war it is huge and floats conspicuously like a blue balloon on the surface of the water. Below the float, in Agalma, there is a series of identical modified medusae that have the sole function of squeezing out water. These muscular structures serve as a means of locomotion and are called the swimming bells. Below these there are clusters and each cluster contains a feeding polyp, male and female reproductive polyps, a protective polyp which has no mouth and contains one long tentacle, and finally a protective modified medusa which is so modified that it possesses little else than a slice of protective jelly. These composite clusters succeed one another so that the lowest one is the oldest, and finally at the very base there is the original primary feeding polyp. This is the direct basal opening of the central gut canal from which all the polyps and medusae branch, and this central canal leads all the way from the basal primary mouth up to the apical float.

Each little Siphonophore is a floating city of quite fantastic complexity. It is integrated as one unit yet basically it is made up of many individuals, and they have become so harmonized in this community that their identity as polyps and medusae is all but lost.

There is one more group related to the hydroids that should be mentioned for not only are they vastly important on the face of the earth but they have one lesson for us here. They are the corals, polyps which build houses, and as is well known these deposits of calcium carbonate which each polyp lays down about itself can accumulate over the years to form great masses of rock and rise above the sea as atolls. Here again we have a case where the communal animal is to some extent bound by his works, for this house governs and confines his activities. Like the nest of a termite it provides a fortress and not only helps keep the enemies without but helps keep the social structure within.

CELL COLONIES

There are many animals that form colonies, but some of these, instead of being a group of multi-celled animals like polyps, may be simply a group of associated single cells. Now it must be admitted immediately that the word "cell" is used here to include quite a wide variety of structures. What is meant is simply one sac, one container, one cell membrane, and within that membrane there would ordinarily be but one nucleus, although in some cases more than one, and the duplication of nuclei takes on various forms. It is striking that among mammals and insects social existence arose independently many times, and now among cells we find the same thing. There are many different types of cell colonies, and it is clear that they have different origins. The colonization of cells was tried many times during the course of evolution and each attempt has special characteristics of its own. These characteristics are to some extent determined by the kinds of cells involved, so let us now take a rapid glance at four different sorts of cells (Fig. 14).

The flagellate cell is probably very primitive, the ancestor of all our higher plants and animals. It has long whip-like flagella, usually one or two, and with these it swims jerkily through the water. Some flagellates are green and have the ability to use the sun's energy to make their food, as do higher plants, while other colorless species absorb liquid food directly or engulf particles like true animals. The colorless line of flagellates has led to the amoeba, a cell devoid of any flagella or consistent cell membrane shape. Instead the amoeba moves by allowing its protoplasm to flow outward into so-called false feet, and in this way manages to pour itself from one place to another. It can feed by absorbing food directly through its membrane, but relies mainly on surrounding and trapping food with its false feet, in that way bringing the food into the center of its body in a

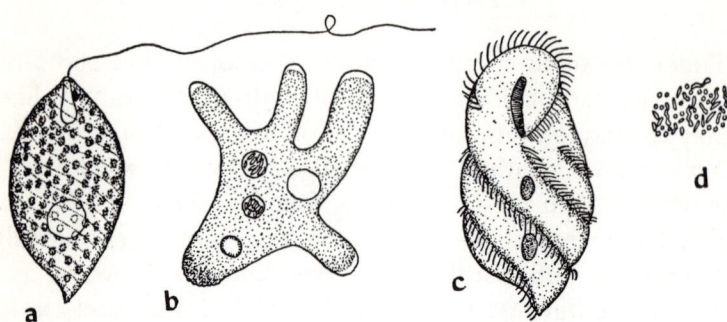

FIG. 14. Four cell types: *a*, a flagellate; *b*, an amoeba; *c*, a ciliate; *d*, different types of bacteria.

small bubble or vacuole. Then its digestive juices draw out the nutriment, and the animal finally ejects the vacuole with the unpalatable remains.

Other relatives of colorless flagellates are the ciliates. These beautiful and complex animals have many whip-like projections, or cilia, and these cilia may be distributed over the body in varied and often intricate patterns. The ciliates also have a mouth, usually a funnel-like structure lined with cilia that waft particles of food inward, and when they collect at the bottom they become detached in a vacuole which, as in the amoeba, floats freely within the internal protoplasm.

Still another cell type, whose relationship in evolution is obscure, is that of the bacteria. Bacterial cells are in many ways quite different from those already mentioned. In the first place bacteria are generally very much smaller than the others and their cell walls are quite rigid, either in the shape of a rod, a sphere, or a corkscrew. They have nuclei but to what extent these minute bodies are equivalent to the nuclei of higher forms is still a matter of active investigation and debate. Bacteria are so small that even their method of locomotion is not clearly known. Some have flagella but a few scientists believe that these are not used in all cases for locomotion but that the organism actually wriggles along by bending movements. There are others which show no wriggling and have no visible flagella yet in some mysterious way they glide along, often at a fair rate considering their small size. As we shall see later, the method of feeding of bacteria is extremely varied, but it is enough to say here that the majority consume dissolved food by direct absorption, while some imitate the method of higher plants and use an outside source of energy to make their own food.

One of the simplest and most obvious kinds of colonies is that of the colonial ciliates, which serve as a good starting point because in many ways they parallel so closely the colonial hydroids. Certain ciliates have an adhesive foot, a supporting stalk which they attach firmly to the bottom of the pond or ocean (for there are both salt and fresh water species). In some cases the individuals are always single, but in others, when the top cell divides longitudinally, the daughter cells remain attached and a branch arises in the adhesive stalk so that the two individuals are now connected to one forked adhesive stalk. By repeated branching a whole tree-like colony will be produced, with the living ciliate cells at the tip of each branch, connected by the branching adhesive stalk laid down behind. This is their means of asexual reproduction or growth, and as in the comparison of isolated hydra with the colonial Obelia, the daughter cells of colonial ciliates do not separate while in isolated ciliates they do. Also as in Obelia, the pattern of the branching is a permanent record of their growth activity, and some species differ from others in their relative growth rates so that the variety of colony shapes is considerable.

A Zoothamnium colony, for example, has the shape of a Christmas tree, one central spire with alternating branches to the right and the left. Each branch has in turn a series of alternating cells, and the lower or older branches have a few cells, the middle section the greatest number which tapers off again to a few at the tip (Fig. 15).

Fig. 15. A colony of Zoothamnium. (From F. M. Summers)

The integration of the colony is dependent upon the type of connections between the cells, though there are rather large differences among species. In some cases the adhesive stalk is merely non-living material exuded by the cells and the cells are therefore quite separate and can be called colonial only because they are stuck to one another. In other cases the adhesive stalk is a hollow cylinder containing in the center a contractile thread of muscle; and in the most advanced forms, such as Zoothamnium, this thread also splits in division and therefore connects all the individuals. This means that a certain degree of communication is possible between cells for here again, if one cell is irritated by poking a needle at it, then with extraordinary rapidity the whole colony pulls itself down close to the base. The impulse to contract is broadcast through the colony by the continuous muscle thread. In some species in which the muscle thread is discontinuous at each branch, the contraction of any individual branch does not affect the rest of the colony.

There is a division of labor in a colony of Zoothamnium. Unlike Obelia, it does not involve feeding, for each cell feeds for itself, but for reproduction there are four different types of cells: at the apex of the colony there is the cell which gives off a new branch each time it divides. There are the terminal branch cells which divide and continually give off branch cells, and there are the branch cells themselves which ordinarily are incapable of division. Then finally there are certain branch cells near the central stalk that enlarge and break off as free-swimming cells that wander away and develop into new colonies. The zoologist Francis Summers showed that if a terminal branch cell is cut off, then another branch cell below it which had ceased to divide will take over the function of dividing and giving off new branch cells. This is also true of the top cell, which will be replaced by one of the cells immediately below it, and every time the new branch cell divides it will give off a new branch. Not only is the labor of reproduction and growth divided, but it is integrated within the whole colony so that there is always only one dominant top cell and one dominant terminal branch cell at the tip of each branch, and this composite picture arises even after mutilation of the colony.

If we now turn to the green or photosynthetic flagellates we find many different kinds of colonial forms, far more than among the colorless cell colonies. The reason for this has been carefully examined by John Baker of Oxford, who suggests that the mode of eating is the source of the difference. For in green forms there is no need for any special feeding apparatus; all that is needed is the presence of the green pigment chlorophyll exposed to the sunlight. Colonial animals, on the other hand, must always have a mouth, a feeding apparatus. They may, as in the colonial ciliates, have a separate mouth for each cell, but obviously this will impose great restrictions on a large colony where a division of the labor of eating would have advantages. But making a communal mouth, such as we find in hydra, is a major engineering problem, a major hurdle in the slow process of evolution, and therefore has been achieved very few times. A good example of a successful solution to the communal feeding problem is the case of sponges. There great masses of cells need food, but only certain flagellated collar cells can capture the food. These cells are in many small hemispherical chambers or along the walls of one large chamber in the less advanced sponges, and they whip the water causing a current to pass through the sponge. The collar cells then grab the particles of food as they pass through, and what they themselves do not eat they give to the many other cells of the sponge. The whole complex canal system allows the sponge, as a community of cells, to get food (and oxygen); it is an elaborate solution to the problem of eating. But plants need only to be green to be nourished and grow, so it is no wonder that we find so many different kinds of plant colonies.

Some of the most interesting examples are among the Volvocales, a group that has a virtually continuous series of forms progressing from single cells to the pinnacle of complexity, Volvox (Fig. 16). A colony of Volvox consists of about two thousand green biflagellated cells that are arranged in a slightly lemon-shaped sphere. Each cell is

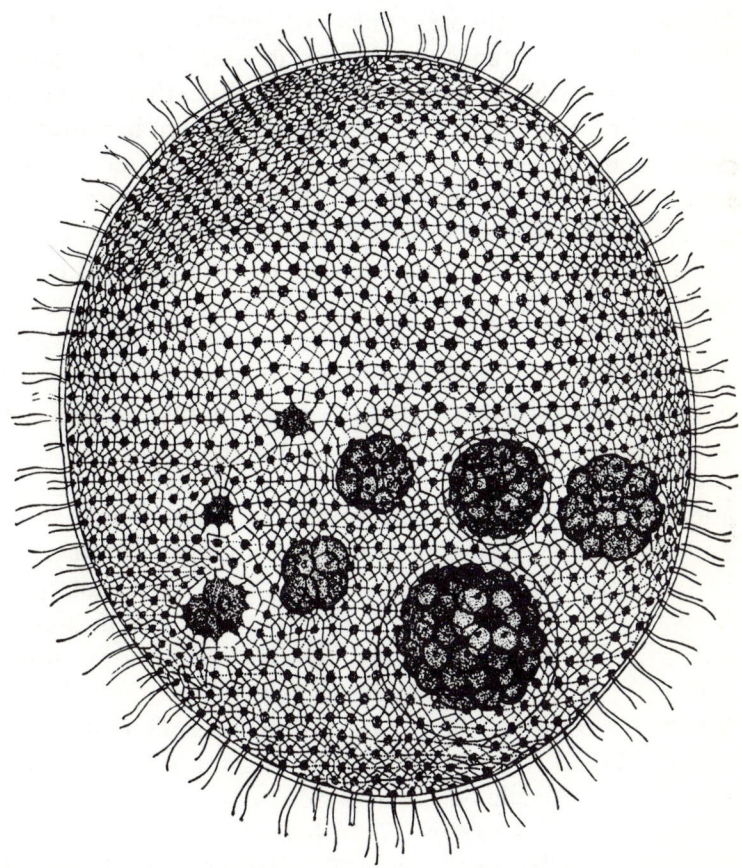

FIG. 16. A colony of Volvox, showing the successive stages in the asexual development of a daughter colony. (From E. G. Conklin)

separated some distance from its neighbor by a matrix of jelly, yet fine protoplasmic strands form connections between each cell in a pattern of hexagons over the surface of the colony. The center of this ball of cells is hollow. The cells which cover the surface show some slight division of labor, some differentiation; and furthermore the whole colony has a definite polarity, a head-and-tailness. The polarity is shown especially well in their movement, for the front end stays upward or forward and the whole tiny globe rotates about a central north-south axis, just as the world does, although Volvox can change the direction of its spinning. It will go merrily in one direction for a while, then suddenly stop and reverse itself. The cells in the northern hemisphere differ from those in the southern hemisphere in two respects. In the first place the northern ones are larger and greener, but it is conceivable that this is merely the result of their being on top and receiving the sun more directly. The second difference is that reproductive cells occur only in the southern and never in the northern hemisphere. There certain vegetative cells begin to swell and may produce sex cells (eggs or sperm), or they may produce a daughter colony directly by asexual reproduction.

But before discussing their reproduction I should like to add a few words about their movement. Even though Volvox is a colony of single flagellated cells, the coordination of their movement is a remarkable thing. They not only keep their front end forward, they not only can in a consistent and precise fashion reverse their direction of swimming, but they can even swim or spin in an oriented way toward a source of light. S. O. Mast of Johns Hopkins University studied this phenomenon and came to the conclusion that the flagella on the dark side gave an uneven stroke, a stronger push away from the body than the pull back, while the flagella on the light side had an even, balanced stroke and in this way the colony is pushed towards the light as it spins. This orientation to light then could be explained by the individual response of the cells depending on whether they were in the light or the dark, while the spinning, its reversal, and the polarity in general seem to be more of an overall communal activity, one which no doubt is greatly helped by the fine protoplasmic connections between the cells.

In the asexual reproduction, as was just said, one of the cells in the southern half enlarges and soon begins to go through a series of repeated divisions. The resulting small cells remain attached at the original site and form a pocket which bulges inward so that eventually there is a daughter colony within the mother colony. This daughter colony then undergoes a remarkable transformation: it turns inside out, just as a sock is turned inside out, at the same time becoming detached from the mother wall and sprouting flagella all over its outside surface. But this free-swimming colony is still within the mother colony and it may not be liberated until the mother dies. Often it is possible to see three generations, one within another. In the sexual reproduction, which comes at the end of the summer, anticipating the rigorous conditions of winter, egg and sperm are produced, and these unite to form a resistant embryo. In the spring this embryo germinates and by a series of divisions forms a new colony which in turn reproduces asexually.

Besides Volvox there is a series of smaller and simpler colonial forms in the Volvocales. Gonium is a flat plate of sixteen cells, Pandorina a small sphere of eight or sixteen cells, and Eudorina is a larger sphere of numerous cells, but not nearly so large as Volvox. Twenty-five years ago a German biologist F. Bock did an interesting series of experiments on the ability of these different colonies to regenerate from isolated cells. He found that one cell removed from Gonium or from Pandorina did regenerate a new colony, but that one cell from Eudorina produced an incomplete abnormal colony. With Volvox, however, as others had found before him, it was impossible to obtain any regeneration at all, showing that as the colonies increased in complexity during the course of evolution, the cells became increasingly interdependent, increasingly rigid in their functions. The direction in evolution appears to have been toward greater unity and integration of the colony, so that finally Volvox almost loses its colonial character and more closely approaches the qualities of an individual organism. But the dividing line between colony and organism is a gradual and tenuous one, well illustrated in the Volvocales series. No doubt an arbitrary definition could be found to distinguish the two clearly, but more important is the fact that the intergrades are numerous and indefinite. Any sharp verbal distinction may be somewhat artificial and misleading.

Let us now turn to the amoeba, a cell type which in a few instances produces a most unusual type of colony, even though the large majority of amoebae are solitary and independent. These colonial forms, the Acrasiales, consist of only eight or so species, all found in moist soil, and we will concentrate on the species Dictyostelium discoideum (Fig. 17).

Dictyostelium will start its cycle if the spores are sown in a sufficiently moist and nutritious environment. In the laboratory a jelly containing dissolved food is used and to it bacteria are added, for the bacteria feed on the nutrient in the jelly, and the amoebae in turn feed on the bacteria. The spores themselves are elliptical, like minute pill capsules and they split down the side to allow the imprisoned amoebae to escape. These free amoebae immediately begin to engulf the bacteria, and as they swell and become replete they approach the period of division. First the nucleus divides in two and the whole cell tears apart into two apparently equal halves, and these halves again eat to repeat the same process until finally in two days time the culture dish is densely populated with amoebae, all direct-

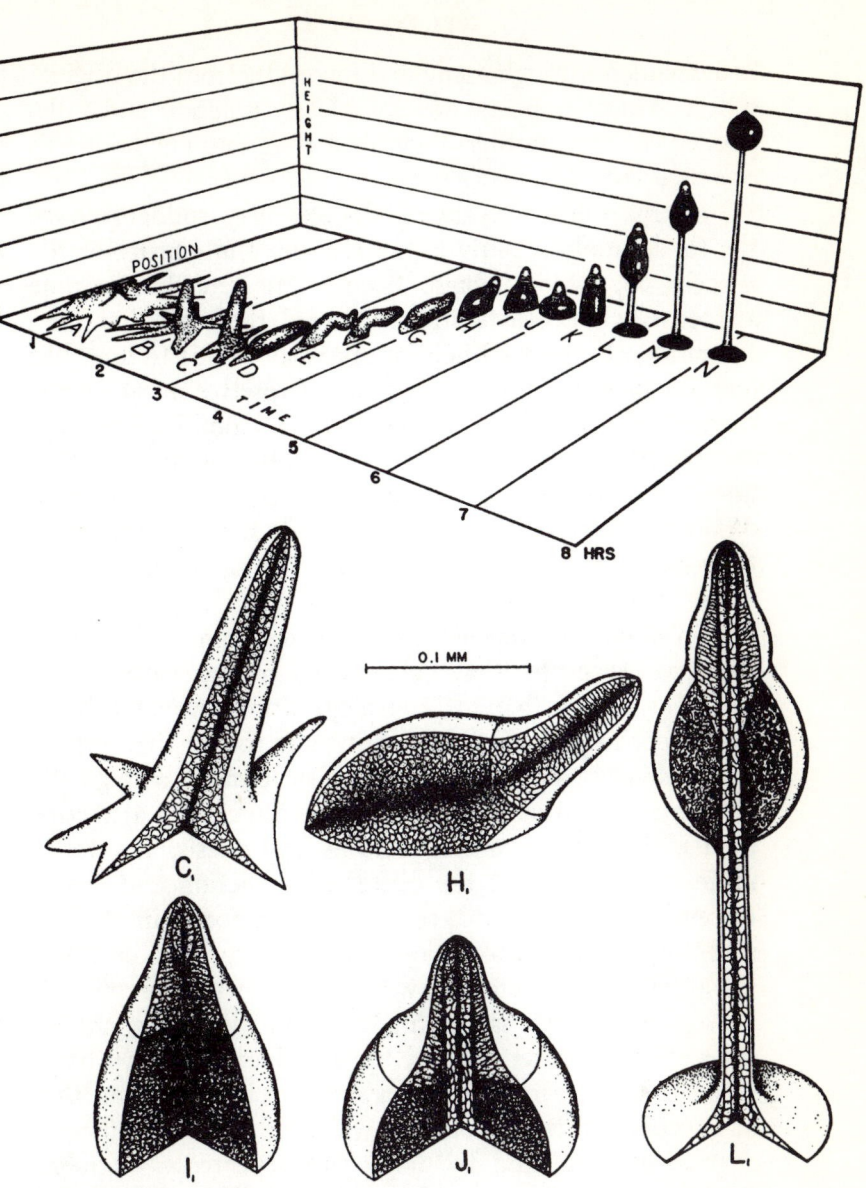

FIG. 17. The life cycle of the cellular slime mold Dictyostelium discoideum. *Above*: The complete development is represented in a three-dimensional graph. A-C, aggregation; D-H, migration; I-N, culmination. *Below*: Semi-diagrammatic drawings showing the cell structure at different stages.

ly descended through many division cycles from the original spores. These amoebae are completely separate and independent of one another; they seem to ignore one another's presence.

But this solitary existence changes very suddenly once the food supply is fairly exhausted and the density of the amoeba population is high. Then certain amoebae become attractive while others stream toward them so that soon the randomly speckled field of amoebae begins to look like feathery crystals formed in a series of rosettes in which the streams of aggregating amoebae are all pouring into the centers. When the process is completed, in two to four hours, all the amoebae will be collected in cell masses of different sizes; and these masses, as we shall see presently, have many communal and integrated properties.

But first it should be mentioned that a great deal is known about the mechanism of aggregation. A few years ago it was suspected that a chemical substance called "acrasin" was responsible for the aggregation. Apparently all the amoebae give off acrasin, but some sooner than others, and perhaps by being first these are surrounded by a greater concentration of acrasin and become the central attraction points. Acrasin, according to this hypothesis, has to be more concentrated at the center of attraction and less so as one radiates outward, so that each amoeba in the surrounding territory will have more acrasin on one end than the other and therefore it will know in what direction to move. We do the same things, although we are notably poor at it compared to, say, the red deer, when we try to locate a smell. First we take a sniff at one point and then at another and try to compare which of the two sniffs gave a stronger odor, and by doing this repeatedly we might find the source. We would be even more like an aggregating amoeba if we had two noses, one at each extremity, and then we could simultaneously make a comparison and in that way guide our movement. Recently B. M. Shaffer of

Cambridge University has confirmed this hypothesis by isolating acrasin and showing that the amoebae will go toward the chemical alone. The transition from the solitary to the social phases of this mold is controlled by a chemical substance. In a sense this was true of certain aspects of some of the higher animal societies we examined, although there the relationships appear to be far more complex. For example, the stags of the red deer are brought into full rut and begin to herd the hinds when the hinds come into heat and give off chemical odors which attract and stimulate the stag. And another example, of the many that were given in earlier chapters, is the chemical substances given off by the termite royal pair which are passed through the colony by licking, and not only prevent future reproductives from arising but also help keep the colony integrated around the central focal point.

In Dictyostelium discoideum the mass of cells that results from the aggregation of the amoebae acquires a sausage shape, which may be from one-half to two millimeters long on the average. This sausage clearly has a front and a hind end, for not only does it move in one direction only, but the front end tapers slightly into a blunt point, and the hind end straggles and leaves behind a collapsed tube of slime, which is in fact a sausage casing. Depending on the environmental conditions, this migrating cell mass may migrate hardly at all, or it may migrate (at the rate of about one millimeter per hour) for one or two weeks before it goes into its final fruiting phase. The migration movement is highly coordinated, and the cell mass acts as one unit. This is shown particularly well in its tendency to migrate toward light and heat. It will go toward an extremely dim light with unerring precision, and even more remarkable, it will move toward a warmer region even when the temperature gradation is ever so small and gradual. In fact it has been calculated that a small cell mass can distinguish a difference in temperature between its two sides of five ten-thousandths of a degree centigrade; the mass as a whole knows which is the warmer and migrates in that direction. This sensitivity seems almost fantastic, and certainly the physics and the chemistry of it are hard to explain.

It is also known that acrasin continues to be produced in the sausage, and furthermore it is produced more at the tip of the sausage than at any other place. It is tempting to speculate whether or not the tip cells (which are the first cells to start the aggregation) are still leading the other cells by attracting them with acrasin. If this is so, it would be like leading a donkey by hanging a carrot in front of his nose, for the tip cells would attract the posterior cells and the posterior cells would as a result push the tip forward. In this hypothesis, then, one would assume that the reason that light or heat affects the direction of movement of the cell mass is because it stimulates greater acrasin production on one side of the tip, giving the tip and that which follows it an orientation with respect to the light or the heat. This is all, of course, pure speculation, and it must be remembered that there is another important factor in the cohesiveness of the cell mass, and that is the stickiness of the cells and the slime sheath or sausage casing, for both these factors tend to keep the cells together and tend to keep them moving in unison in the same direction. The actual movement itself appears to be the result of the individual amoebae within the cell mass which are all sending out false feet and in some way they get traction and force the whole mass to move.

We have so far seen the method of feeding of Dictyostelium and the method of multiplication, as well as some aspects of its coordination when in its colonial phase. Now we come to the most remarkable aspect of all, its division of labor, which becomes evident as it enters the final fruiting phase. During migration the amoebae within the cell mass begin to show differences: the amoebae in approximately the forward third become slightly larger and stain differently with various dyes than the amoebae in the posterior two thirds. The dividing line between these two regions is very sharp, and it is especially interesting that in cell masses of different sizes the proportion of the two cell-types is always constant, showing that there is some sort of regulating mechanism which adjusts the proportion. At the end of migration the elongated sausage which, in the laboratory, has been crawling about the surface of the nutrient jelly, contracts into a ball and the tip assumes an upward position. Some of the larger tip cells now suddenly become greatly enlarged and push downward like a wedge through the smaller posterior cells. It is evident that this is the beginning of a stalk, for once this cylinder of swollen cells has reached the bottom, the other tip cells begin piling on top to build the stalk up at the top end. The result is that as the tip cells become used up an ever-elongated stalk, a delicately tapered cylinder of cellulose packed with large cells, rises upward into the air, often to a height of two or three millimeters. The smaller posterior cells are lifted up, like a bag, or perhaps more like a tiny drop of water, into the air so that they always surround the rising tip. This sack of cells will be the spores, and during the rise each amoeba in the sack becomes condensed and surrounded by the hard wall of the spore capsule. Here, then, is a true division of labor, for some cells (the forward ones) become imprisoned in the stalk, while the others (the posterior ones) become spores and are capable of perpetuating the species.

That it is not known how the cells are divided in such a precise and proportionate manner in the migrating sausage is not surprising, for this differentiation is part of the fundamental riddle of the development of all living organisms. But we know some facts which help to reveal the character of this division of labor. If a sausage is cut across the middle, which can easily be done in the laboratory, each section will become a complete individual with anterior stalk cells and posterior spore cells. Furthermore a sausage will sometimes split lengthwise of its own accord and each half will produce a normal fruiting body. Also upon occasion two masses may fuse if they bump into one another and are going in the same direction, and this will produce one large, normal individual. In developing into a fruiting structure, the organism somehow takes cognizance of its size, and accordingly produces a proportionate final form. So there is not only a division of labor here, but it is controlled in a rigid and integrated fashion.

In discussing other types of cell colonies it was mentioned that the problem of feeding animal colonies was greater than that of feeding green photosynthetic plant colonies, and Dictyostelium is particularly interesting in this respect. For here the problem is really avoided rather than met head on; instead of producing a communal mouth, the amoebae feed while they are still separate and independent, and all the remaining stages of the life cycle, where the cells come together as a colony, are done without feeding, using the energy previously accumulated during the feeding stage. But the fact that the colonial stages take place without feeding does impose rigid and crippling conditions upon those stages, for they are strictly limited in the amount of energy they can expend. As a result they soon arrive at a condition of rest, of dormancy, that is the final fruiting body in which the stalk cells have become essentially dead, trapped in the congealed cellulose walls, and the spores lie in hibernation, waiting for favorable conditions in which they can again feed as separate amoebae. In some superficial ways this parallels the fur seal bulls, the harem masters who separate in time their feeding activities and their community activities. They feed as solitary individuals for the long winter months, and for the two summer months they fast and keep harems, entirely on the energy in the form of fat that they accumulated during the winter.

There is another kind of colony that closely resembles Dictyostelium, but instead of amoebae it is made up of bacteria cells. The so-called slime bacteria consist of about two dozen species, but most interesting is the large and elaborate Chondromyces (Fig. 18).

In Chondromyces the lemon-shaped orange-yellow spores, or cysts as they are more properly called, do not consist of

FIG. 18. A mature fruiting structure of the slime bacterium Chondromyces showing the clusters of apical cysts at the tips of the exuded stalk. (From R. Thaxter)

one bacterium but of several thousand. Upon germination they split open and the rod-shaped bacteria stream out like a flame from the mouth of a dragon. From the very beginning these rods, which move in some mysterious way without the aid of any false feet or flagella, come together in masses; at all times they appear to be intensely gregarious.

This means that the mass from one cyst will soon join another, often a larger one, so that as the growth phase progresses the size of the masses becomes increasingly greater, for there is a constant fusion and attraction of the smaller masses. During this period the rods themselves feed and divide; they absorb dissolved substances directly from their surroundings, so that the masses not only increase in size by fusion, but also by the direct multiplication of the rods. The large masses themselves move or more properly pour first in one direction and then the other, and the larger they become, the more impressive are their movements. Sometimes the rods seem to fan out, as though in search of fresh food, and then they come together again in a small knob.

The mechanism of orientation or coordination of this movement is in itself an interesting problem. There appear to be two agencies at work: for one the rods tend to move along the pathways where previous rods have been; apparently each rod leaves behind a track that the other rods follow. The other factor is demonstrated when there are no tracks between a large mass and a small group of rods for then the rods will somehow be attracted at a distance to the larger mass, indicating that perhaps here also, as in Dictyostelium, there may be some sort of chemical attraction, but the evidence in Chondromyces is completely lacking. In observing the movement of the rods of this slime bacterium, one is often reminded of the marching columns of army ants for there, as we saw, one ant will slavishly follow the tracks of another. In Chondromyces this follow-the-leader process can occur because there are actual grooves which the previous rods laid down that can be followed, while in the ants it is known to be a chemical scent left on the trail. The result, however, is that many of the marching patterns of ants and Chondromyces are similar, and just as circular marches have been observed in army ants, they also exist in the bacterial rods, although the rods seem eventually to be able to break loose and wander off, while the army ants will usually march in circles until they die.

When the masses of rods become large enough the final fruiting may begin. This is unfortunately not quite so interesting as in Dictyostelium as there is no real division of labor, although the fruiting structure itself is certainly elaborate. The heaps of rods seem to rise into the moist air mainly by producing great quantities of a non-cellular slime that accumulates behind them. In this way they may rise a good millimeter into the air, the rods always in a group at the tip, but this tip occasionally divides in two so that the final structure will be branched, the stalk always consisting of clear yellowish slime. When the rods reach their final height they suddenly become carved out into many small spheres and each one of these spheres is a deep yellow cyst. Nearly all the rods enter cysts, and only a few are left straggling in the completed stalk. It is noteworthy that cells as low in the evolutionary scale as bacteria can have such a complex type of colonial existence, even though they lack a true division of labor.

In the four examples of cell colonies that we have seen it is unequivocally clear that each attempt at colonization must have been quite independent, for the ciliate cell, the green flagellate, the amoeba, and the bacterium are so entirely different. Yet despite the separate origins the need for feeding, for reproduction, and for coordination is the same, and therefore it is not surprising to find many similarities as well as differences. For instance in each case the method of locomotion is different, but in Dictyostelium and Chondromyces there are aggregations of a quite similar nature. If the reproduction of the four types is compared we see that in Zoothamnium and Volvox a new colony is produced by a single cell which divides many times and the daughter cells cannot detach themselves, while in Dictyostelium and Chondromyces the division occurs in the phase where the cells are relatively separate, and they are brought together by an aggregating mechanism. But all these similarities and diversities are variations within strict confines, the same strict confines that affected mammal and insect societies, the need of a living unit to perform certain specific living activities.

12

A SINGLE CELL

WE HAVE slowly steered our course toward more and more simple colonies or groupings, going all the way from complex mammal societies to societies of single cells. Now we can take still another step in the same direction and examine what might be called living colonies of molecules, that is, a single cell. For each isolated living cell is a community of chemical substances existing together, and as in all the higher groupings, they must eat and obtain energy, have some coordination and even division of labor, and they must reproduce. But again it should be emphasized that it is not to be expected that the molecules within a cell will perform these functions like monkeys in a monkey clan; it is simply that the functions are the same though they are performed by different groups in different ways.

As we saw previously there are so many different kinds of single-celled organisms, and some, such as bacteria, are so radically different that it is impossible to choose a representative type. Instead the most complex type will be chosen, for the complexity itself is interesting and shows to what degree a single cell may become elaborated. A ciliate best fits this description, and ciliates are particle-feeders. This means that the green photosynthetic single cells and the various types of bacterial cells will be neglected for the moment, but their activities, most especially their feeding mechanisms, will be touched upon in later chapters.

The ciliate Euplotes is a small animal about one tenth of a millimeter long, found sometimes in abundance in rather stagnant, foul water (either fresh or salt depending on the species) (Fig. 19). It is shaped like a slightly cupped hand with the fingers tight together; the back of Euplotes

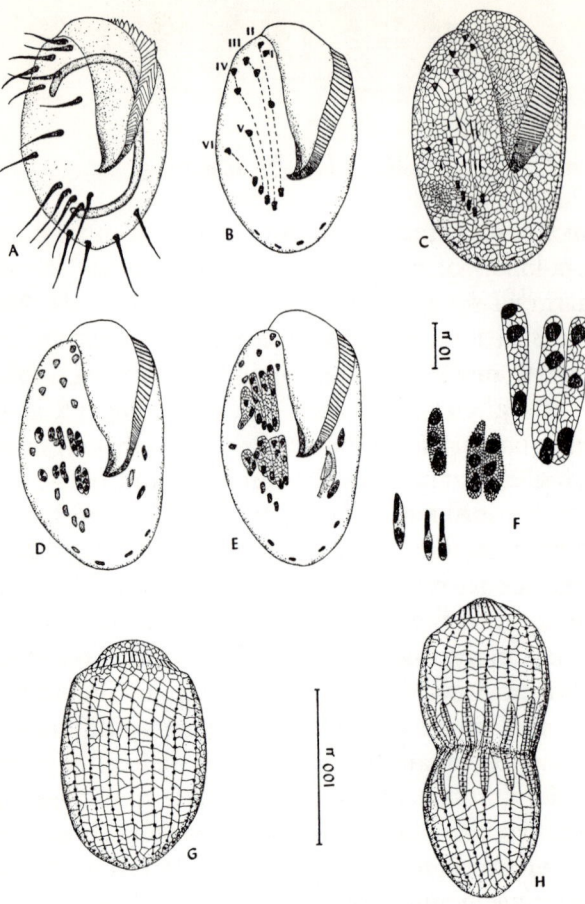

FIG. 19. The ciliate Euplotes. A, ventral view showing the cirri, the mouth structures and the large C-shaped nucleus; B, ventral view indicating (by dotted lines) the cirri that originate from the same group; C, silver-stained preparations showing the silver network and the beginning of the primordia of the new cirri anticipating cell division; D, E, showing the progressive stages of the cirrus development prior to division; F, three successive stages of cirri expansion (from lower left to upper right) showing the progressive widening of the spaces between the cirri; G, dorsal view showing the rows of basal bodies which lie at the base of the bristles; H, a similar view during division (note the multiplication of the basal bodies near the equatorial division zone).

is curved while the front or ventral side is irregular and flat. The mouth cuts the ventral surface in a conical groove like a miniature whirlpool, and along the margin of the mouth, extending up over the front end, is a band of thin transparent membranes arranged in parallel rows (like the broad teeth of a comb), that are constantly moving back and forth in rhythmical unison. The rounded back side has seven rows of bristles that run, like the longitude lines on a globe, from the north to the south pole, but on the ventral side there are eighteen irregularly placed large cirri, which are cilia plastered together to form one bristle just as the hairs in a camel's hair paint brush fuse when wet to form one large point. These cirri are arranged in a very specific pattern, consistent in each species.

Inside the cell there is a large nucleus which is shaped like a long sausage but bent in the form of a "C." There is also a small nucleus lying just behind the back of the "C," and the functions of these two nuclei will be discussed presently. Besides the nuclei there are many food vacuoles inside the protoplasm, and also numerous small bodies of various sorts, some plastered underneath the surface, all over the body. By using various chemical stains it is possible under the microscope to reveal that there are fibers at the base of each cirrus which in most cases give the appearance of short whiskers, but each of the five large anal cirri has one fiber that extends straight up to the front end of the animal like the strings of a violin. By using a special silver-staining technique it was found that the whole surface is covered with a lattice-work like the veins on a leaf, and the dorsal bristles as well as the ventral cirri are all attached to this lattice network. It is also possible to see that in the lower ventral region there is a small opening surrounded by a rather dense lattice; this is the opening of the contractile vacuole.

In the microscopic jungle of one-celled creatures Euplotes is a fierce predator which feeds on smaller animals such as flagellates. The delicate membranes that surround the mouth of Euplotes wave rhythmically and set up water currents which literally suck the food down into the whirl-pool-shaped mouth. As soon as the helpless prey is at the base of the mouth a small bubble is blown into the internal protoplasm and in this bubble of water lies the victim. The bubble-vacuole then becomes detached and floats inward into the protoplasm. For a while the small flagellate will jerk about in its prison, but soon the coagulative digestive juices will seep into the vacuole and paralyze him, and then the digestible material of his body will become liquefied and it in turn will seep back into the living protoplasm of the Euplotes. Finally all that remains of the flagellate is the empty carcass, the unpalatable remains, and this, still in its vacuole, will move toward the surface where the bubble will burst to the outside, ejecting the carcass as excreta.

If food is abundant the body of the Euplotes may be literally crammed with food vacuoles, and since the process of eating and digestion involves a great intake of water, there must be some mechanism to provide a water balance so that the whole animal will not burst.

In Euplotes the water is removed by a contractile vacuole. This structure, like a food vacuole, is a bubble that forms in the lower part of the body, and it slowly swells as the excess water from the body pours into it. When it reaches a large size, this clear bubble suddenly bursts and disappears, and the water is ejected to the outside through the special pore at the surface on the lower ventral side. The bubble then forms again and by rhythmic contractions and expansions the bailing out takes place. In this way the cell can get rid of certain soluble waste materials as well as water (what amounts to urine) and at the same time contain digestive juices, salts, sugars, and many other dissolved substances in a concentration higher than the environment. In order to concentrate the molecules necessary for life in one package and at the same time have some flexibility and sensitivity to the outside confining membrane this contractile vacuole mechanism is very cleverly devised. The importance of having a delicate membrane (at least in some regions) is clear enough in Euplotes, for how else could it take in particles of food and eject the excreta? Eating particles rather than dissolved food requires some such mechanism as the contractile vacuole, unless the cell lives in an environment where the environment itself has many dissolved substances in it, such as the sea, and it is true that many marine cells lack contractile vacuoles. We see here, in the matter of the water balance of cells, a problem associated with feeding and maintenance that has no parallel in the higher animal societies. It is a problem that is purely physical in nature, that exists only in cells and groups of cells, but not between individuals.

A few minutes' observation of Euplotes under a low-power microscope is all that is necessary to convince one that its movements are highly coordinated. Not only do the rows of membranes about the mouth beat in unison and manage to waft food down the gullet, but the movements of the whole animal depend entirely on the activity of the cirri. In the first place the fact that the ventral surface possesses all the cirri (the short dorsal bristles do not appear to be involved in locomotion) means that if it moves on a surface, such as the bottom of a glass dish, it walks on its cirri as though they were legs. It can also swim freely through the water, front-end always forward, or it may come up and lie on its back just under the surface, its cirri hanging onto the air-water interface. It is not only that it can, with its eighteen cirri, perform these different feats, but it does them with a professional assurance right from the moment it is born. If one examines the movement of the individual cirri it can be seen that sometimes they all move in unison, sometimes they all stop, and at other moments some move while others remain stationary. They may move slowly or they may move fast, and among the eighteen cirri the overall motion is smooth and perfectly coordinated.

One question that has always intrigued biologists is whether the fibers which attach to the bases of the cirri are involved in nervous coordination. They look so much like nerve fibers that the analogy is tempting, and furthermore, C. V. Taylor of Stanford University showed that if he made incisions so that the fibers were cut, the five cirri involved continued to move, but their movements were quite random and disrupted. But unfortunately this still does not prove their coordinating function, for these animals are so small that any such incision, no matter how carefully made, is bound to disrupt a considerable area, and the cause of the lack of the coordination in the cirri could as well be disturbances other than the cutting of the fibers. Unfortunately all that can be said at the moment is that there is nothing to oppose the idea that the fibers are responsible for carrying messages from one group of cirri to another, but there is no positive assurance that this is their function.

Relative to the matter of coordination, there is a high degree of division of labor in Euplotes, and it is especially interesting that the division of labor is within one cell. Not only does the animal have a front and a hind end as well as a belly and a back, but it has a mouth with its attendant structures, a contractile vacuole with a special ejection pore, dorsal bristles and ventral cirri. All these structures have special functions; in specific ways they are associated with feeding, with water balance and excretion, or with locomotion.

Before coming to the details of reproduction in Euplotes it may be well to say something about cell division in general, for Euplotes is a rather special case. In the first place most cells have but one nucleus and this nucleus is made up of a number of thread-like bodies called chromosomes. On the chromosomes there are genes and these genes are responsible, at least to a great extent, for the character of the surrounding protoplasm. They are the units of heredity and presumably the vast majority of characters which an individual of any sort possesses—size, shape, color, etc.—are determined by the genes. During development the genes actually govern and control the activities of the outside protoplasm or cytoplasm. There is some evidence for additional control factors carried in the cytoplasm, but the nucleus with its gene-bearing chromosomes carries the brunt of the task.

Before cell division, in a resting nucleus, the chromosomes usually cannot be seen, but as division approaches they condense and sort out, so that they look like strands of spaghetti, ranging in number depending on the species from two or three to many hundreds. Their next step is to divide so that instead of being single strands they are double. Following this they become arranged in a flat plate across the cell, and a "spindle" forms about them, with fine spindle fibers attached to the chromosomes. These spindles appear to pull at the chromosomes and one of each pair goes to an opposite pole so that there is a complete, equal division of the gene material. As the two sets of daughter chromosomes go to the poles they again become transformed into two resting nuclei, similar to but smaller than the original parent. As the nuclei pull apart the cytoplasm divides in two; in some forms it pinches apart in the center, and in forms with rigid cell walls a dividing plate is formed separating the two cells.

In Euplotes, where there are two nuclei and so many structures covering the surface, cell division is a bit more complicated. As far as the nuclei are concerned, only the small nucleus divides in the way described above, and the sausage-shaped large nucleus simply pinches in two as the whole cell pinches in two. A number of investigators have studied the relative rôle of the two nuclei, especially in other related ciliates, and it is now agreed that the large

nucleus is responsible for the production of the new cytoplasmic structures after division. This is shown especially
clearly in regeneration, where new structures can be produced even in the absence of a small nucleus. The small
nucleus is involved in sex mating and can pass and mix the
characters of one individual with another. During sexual
union the large nucleus disappears, and after the mates
separate it is reconstituted, but the new large nucleus acquires the characters of the newly crossed small nucleus.
So the small nucleus governs the character of the large
nucleus which in turn governs the character of the cell.
This phenomenon was shown especially well by some ingenious experiments of Tracy Sonneborn of Indiana University, who produced individuals with a different type of
large and small nucleus, and the cytoplasm always developed
according to the direction of the large one and never of
the small.

The division of the cytoplasm in Euplotes is most complex and curious. In the first place the anterior daughter
cell keeps the original parental mouth, while the posterior
cell produces a new one. On the back side of the cell the
bristles divide mainly before the cell divides, in preparation
for the oncoming event. The cirri on the ventral side are
completely rebuilt, that is, the old ones are reabsorbed and
new ones form in a small bud for each daughter cell and
they spread out and assume their correct position as division proceeds.

I have already mentioned that there is a sexual mating in
Euplotes. This is not strictly speaking a reproduction, for
two individuals of opposite sex or mating types come together and form a protoplasmic bridge that connects the
two. After a complex series of nuclear events they fuse their
micronuclei and then separate so that although they have
not produced offspring, each parent now has, like an offspring, a new combined set of genes (half from its former
self and half from its mate). These altered or offspring-like

parents now divide asexually and the subsequent generations have the new gene combination.

There are many lessons to learn from Euplotes such as the fact that a single cell is not necessarily simple, and quite to the contrary it is amazing that so many complicated structures and activities could all be packed into one cell. One must admire it the way one admires a minute Swiss watch which may keep time as well as the grandfather clock in the hall. In Euplotes it is not a matter of keeping time, but of feeding, and reproducing, and performing all its activities in a coordinated fashion, and this it does, as we have seen, with ability and skill.

13

ENERGY, MATTER AND CELLS

THE TIME has come, now that we have reached the bottom of our downward exploration and have examined a single-celled animal, to ask just what is really meant by feeding, reproduction, and coordination. We say, for instance, that an animal must eat to live, and we all know this from personal experience, but now let us ask why this is so. The answers to these problems lie not in biology but in the province of physics and chemistry, so we shall linger there for a brief instant.

It is important to keep in mind that each of these problems—the physical and chemical basis of feeding, reproduction, and coordination—has perfect examples in the non-living world, especially the world of machines and man-made devices. They are not problems peculiar to the living alone, and any individual activity or manifestation of life can usually be imitated by a mechanical device; the special nature of living things comes from the particular assemblage of specific kinds of matter in such a specific way that a large variety of activities can be performed. I admit that this is rather a mechanistic view of life, but it has its compensation in its simplicity. At least for the sake of argument let us consider living things as machines, albeit specially made and beautifully designed, and that our interest as scientists and human beings is to understand more clearly how they are put together.

If we turn now to feeding, the reason animals must eat to live is the same reason cars must have gasoline to move. The fuel in the car is a source of energy, and the car can use this energy by taking it into its cylinders and exploding it there. Most of the so-called chemical energy of the fuel is converted to the mechanical energy of moving the piston and the car, but the efficiency of this process is hardly perfect, and some of the chemical energy is lost in the form of heat, for as we all know, the engine of our car gets extremely hot and has to be specially cooled.

However, as the plant physiologist David Goddard has pointed out, there is one big difference between a living organism and any machine, and this is that a machine is made of special noncombustible materials such as steel, while a living organism is made of the same material it consumes and burns. Imagine a wood furnace built of wood— how long would it last?—yet this is exactly the same as when you and I eat beefsteak. This peculiar state of affairs is possible because the actual combustion of the food takes place within the cells (cellular respiration as it is called) at such a slow rate that only a slight warmth comes from it, not enough to set the cell and the rest of the body on fire. By using the energy slowly a living cell can exist and can benefit from the energy for a long period of time. The enzymes or protoplasmic catalysts are responsible for this slow burning, but how they perform this delicate task must wait for an instant while we discuss what is meant by combustion.

Combustion or oxidation used to be thought to involve the addition of oxygen to a substance (hence the word "oxidation") as is involved when the carbon of burning coal combines with oxygen to form carbon monoxide and then by another similar step carbon dioxide. It was then realized that some reactions such as the combination of carbon and fluorine gas "burned" extremely vigorously in the absence of oxygen. So it was realized that this important energy-releasing process of combustion was really a specific kind of internal change in the molecules of the substance, and oxygen or some other chemical could mediate this change. Most combustion that takes place within living organisms involves oxygen, and we call this aerobic respiration; but as Louis Pasteur showed long ago, many organisms, especially some bacteria and yeast, can live and eat in the absence of oxygen; they oxidize their sugar without the use of oxygen. This is in fact fermentation, which Pasteur called "la vie sans air," and it is a common fact that alcohol can be produced from sugar by yeast only in the absence of oxygen; if air is admitted the yeast makes vinegar instead. In other words, the yeast eats the grape sugar and if air is absent the chemical reaction runs in such a way that the final product, after the free energy has been removed, is alcohol. But if oxygen is present another easier chemical reaction takes place to exploit the free energy and acetic acid is produced instead of alcohol.

When food (carbohydrates, fats, or proteins) enters a cell it is not completely oxidized in one or two chemical reactions (as happens in a coal furnace) but the process requires many steps, and a most important fact is that each step is controlled by an enzyme. Enzymes are proteins, large complex molecules, and each has a specific task. One enzyme will combine with a particular substance and only with that substance, and during the course of the combination the substance will be slowly degraded into an altered substance or usually two substances, and each of these will be attacked by another specific enzyme which in turn will slowly break down the product into again other substances. In each step energy is released, and this energy may be used to perform the functions and activities of the cells, or the energy may be lost in the form of heat.

Biochemists now know a great many of these steps and the enzymes involved; this research has been one of the important advances in the last fifty years of biology. It is now realized that the fuel of the living cell can be the same material as the cell itself because these batteries of enzymes are perfectly designed to slow combustion to a snail's pace. It is possible, by using various kinds of sensitive pressure- or volume-changing gauges, to record the combustion or respiration of cells, and it has been observed that if a piece of tissue is macerated brutally it gives a sudden great increase in respiration before it dies. This gives a further clue to the slowing process that has nothing to do with the pure chemistry of the reaction; it shows that not only is it necessary to have certain enzymes in a cell, but their spatial configuration is vitally important. The degradation of the high-energy food can only follow certain pathways because the enzymes are placed within the cell in such a way that this one slow pathway is possible, and any maceration disrupts this order and the whole system is short circuited when the enzymes are mixed up. After years of research we still do not know much about the actual arrangement of enzymes within a cell, but it is a subject of intense interest and active research.

Some of the energy released during respiration (excluding that which is lost in the form of heat) is used as in the automobile directly in movement. In animals, for instance, where there is muscle which can contract, this contraction process is energy-consuming. Muscle itself involves contractile protein molecules and these molecules need to be activated, that is, charged with energy, for each contraction. The process is a purely chemical one, again involving numerous steps, and each time the contractile protein has lost its energy it is again recharged by energy which has been brought through channels from the original high-energy food.

Another use of the energy released by combustion within the cell leads us directly into the problem of the physical and chemical basis of growth and respiration, for some of the energy is used in the synthesis of new protoplasm. To make carbohydrates, fats, proteins, and all the other complex molecules characteristic of protoplasm it is necessary to have energy, for this synthesis is an uphill process. So by taking the energy from the degradation of food it is possible, again in a long series of enzyme-controlled steps, to make new proteins and other substances. This is exactly what occurs in growth and even occurs constantly in a fully grown mature individual as the various substances in the body are constantly replaced by new ones. The chemistry of growth involves the series of chemical reactions which produce new protoplasmic substances, and these chemical reactions are possible only because, by eating, the organism gets energy to push along these synthetic processes.

Of special interest is the kind of growth seen in complex proteins and nucleo-proteins where molecules produce, with the help of a source of energy, more molecules identical to themselves. During cell division, when a chromosome becomes double and the genes also have doubled, they have reproduced themselves in such a way that all the complex chemical structure has been duplicated. Biochemists have called this kind of reproduction a template process, for one group of molecules seem to be able to mold another identical to themselves, no matter how complex their structures. The reproduction of virus involves the same process, for a few molecules of virus protein in a living cell will, sometimes in a few hours, produce millions of molecules identical to the original ones. For both the gene and the virus it must be remembered that this process can take place only with an outside source of free energy, and somehow a living cell can produce the energy necessary for this remarkable kind of template reproduction.

It is harder to talk of coordination within a cell on a molecular level. Obviously there is among molecules a division of labor, and the enzymes perform different functions from the food sugar, but this seems almost too remote from the sort of division of labor that was involved in the gross morphological structures of cells and groups of cells. There is also between parts a chemical communication, and the reactions take certain specific pathways, but again the analogy is remote. In the chemistry and physics of the conduction of impulses along a nerve there is a real molecular basis for living communication, and all cells, nerves or otherwise, are capable of transmitting impulses over their surfaces.

In all these functions of living cells it is clearly necessary to constantly pour in free energy, for the motor is never quiet except in death, and as the motor turns, energy is constantly channeled into different pursuits. Life is a steady state in which the intake of energy is always balanced with the energy loss, and between this inlet and outlet of energy turn the wheels of all the living activities.

14

FEEDING IN PLANTS

In the beginning of this book various animal societies were used to stress the importance of certain living activities. We have followed these same activities through lower colonial organisms, through single-celled organisms, and we have even penetrated into physics and chemistry to explore the basis of these living functions. Now armed with this background it is possible to weave back up the evolutionary scale toward higher forms and see how these living activities operate within multicellular organisms, to understand the fundamentals of these activities which are so characteristic of any living unit. But in their evolution organisms have made one major split aside from countless minor bifurcations, and we have the greatly divergent kingdoms of plants and animals. In order to simplify the presentation we will first examine plants and begin with the matter of intake of energy by plants.

Before discussing the green plants, which get their energy by photosynthesis from the sun, we shall look at some species of bacteria that have an extraordinary ability to obtain energy from an extraneous chemical reaction which they encourage, and with this energy they perform all their life functions. These so-called chemosynthetic bacteria vary in the kind of chemical reaction they can stimulate, but in each case the method is basically similar. The bacteria are little salesmen in that they promote the reaction of buying and from this they grab their commission, the energy upon which they live.

A good example is the bacteria that live in the soil and burn ammonia. That is, they encourage the combination of ammonia and oxygen to give nitrous acid and water, and this reaction liberates energy which they capture and use to convert carbon dioxide and water into sugars. This is an uphill reaction, from water and carbon dioxide to sugar, but with a good measure of energy the push up the hill is sufficient, and now in the form of sugar the bacterium has real high-energy fuel. By burning this sugar in respiration it can in turn get energy to build its proteins and all the other substances it needs; its minute machine can run smoothly provided it has sugar to burn, and it makes its own sugar by promoting and parasitizing the oxidation of ammonia. These bacteria are incidentally very important in the soil for they fix the gaseous nitrogen of ammonia into nitrogen compounds, so necessary in the fertility of the land.

Ammonia is not the only substance attacked. For example another species will turn nitrous acid into nitric acid, and again energy is captured. In many swamps, especially where the mud is black, hydrogen sulphide is burned to sulphur by chemosynthetic bacteria, and again sulphur may be oxidized to sulphuric acid, and in each reaction the bacterium gains energy. The iron bacteria provide another example, one which is of considerable nuisance value, for they encourage the rusting of iron pipes. Ferrous carbonate combines with water and oxygen to form ferric hydroxide or rust, and here the gain of the bacterium is the loss of the water company. Sometimes simple organic compounds are attacked, and another swamp species of bacterium oxidizes methane gas to get its energy. But perhaps the most remarkable of all is the species which burns hydrogen gas, giving water, and from the energy derived from this simple combination it can build up all the complicated chemical constituents of protoplasm.

All that is required, in the case of these chemosynthetic organisms, is the presence of some source of energy, carbon dioxide, and water, and from then on, except for the materials necessary to make various types of internal compounds, they are entirely self-sufficient. But the fact that they must always be near some chemical substance that they can break down to get energy means that their position in the environment is strictly limited, and it is in this respect that photosynthetic plants have a great advantage. An iron bacterium must always be near iron, but a green plant need only be in the light, and light is far more abundant and widespread on the surface of the earth than deposits of iron or sulphur.

Basically the process of photosynthesis is the same as that of chemosynthesis. With energy, carbon dioxide and water combine to produce carbohydrates. The real difference lies in the source of the energy. The green pigment of plants absorbs light; it is hit by light particles and after the collision the chlorophyll molecule becomes violently agitated; it is bumped from a sluggish low-energy state to an active high-energy state. In this condition the chlorophyll can release its energy and force carbon dioxide and water to become carbohydrate and oxygen. The details of the chemical process are extremely complex and involve a number of steps still not wholly understood, but the principle of the process and the end result are crystal clear.

It will be noted that this photosynthesis reaction is the reverse of respiration and the two may be written together in one equation:

$$\text{Carbon Dioxide and Water} \underset{\text{respiration}}{\overset{\text{light-chlorophyll}}{\rightleftharpoons}} \text{Sugar and Oxygen}$$

This means that if the plant is to use sugar as a fuel, which it does constantly so that it can make new substances and keep its machinery going, it will undo the advantage gained by photosynthesis and destroy all the valuable sugars. Obviously, if the plant is to grow, or for that matter exist at all, the total amount of photosynthesis must exceed the respiration so that the plant can accumulate energy. During the night respiration alone takes place and there is considerable degradation; during the day both take place, but photosynthesis outstrips respiration by a great amount and the bank account of fuel is always kept in excellent condition. In fact photosynthesis is so effective that not only does it provide for the plant, but the whole animal world is ultimately dependent on this source of available or "free" energy. The struggle for existence, the physicist Ludwig Boltzman said many years ago, is the struggle for free energy, and the free energy of the whole living world comes almost entirely from photosynthesis, the insignificant group of chemosynthetic bacteria being the exception.

Since animals are dependent on plants for their fuel it is obvious that plants must have arisen first in the early history of the earth. But it is not at all certain whether chemosynthesis or the more successful photosynthesis was the first method of energy capture. Those who favor the chemosynthetic organisms as the earliest living forms point out the extreme simplicity of the types of degradation involved as compared to the complex molecule of chlorophyll necessary for photosynthesis. The opposition point out that chlorophyll is no more complex than many other protoplasmic constituents and furthermore light is so omnipresent that it is much more likely, by odds alone, to have been the father of the first living machines. These events occurred many millions of years past, and all traces have long ago disappeared, so the argument is likely to have no end. The matter is even further complicated by some purple bacteria which C. B. van Neil of Stanford University showed are really mid-way between. They need both light and an extraneous degradation of some simple chemical substance to obtain energy, so perhaps these forms are a common ancestor to both, but the evidence for this suggestion is equally insecure.

Whatever the beginnings, the photosynthetic organisms were soon overwhelmingly successful and evolved into many different forms and shapes, finding their way into every conceivable kind of environment: salt and fresh water, swamp, desert, mountain, plain. The only place they could not enter, at least without modifying their mode of nutrition, were places devoid of light. And with this expansion of habitat came a broadening of their structure; the simple bacterial cells were succeeded by filaments of the algae which in different ways produced larger and larger cell masses, achieving great size in the marine kelp, and then with the reconquering of land of the larger forms came the evolution of all our higher land plants, all the trees small and large, the bushes and the grasses. In every step of this evolution, in every modification of form that arose in the course of natural selection, one factor was never neglected and that is that the plant must capture light as efficiently as possible. This was no problem at all with the minute single-celled algae and even the small colonies such as Volvox, but when the size of the organism and its thickness increased, the problem of getting enough light for photosynthesis became serious and limiting. The problem is easy to visualize in terms of two hypothetical plants shaped like round balls, one a fraction of a millimeter thick and the other the size of a basketball. The sun striking on these green balls will penetrate to all the cells in the small one, but only a small proportion of cells in the larger one will get enough light, just the cells on the outside layer. So if plants are to compete successfully with their neighbors and still retain a large size, they must devise more and more efficient ways of capturing sunlight.

The obvious course is to produce structures that are not spherical balls, but that are flat and thin so that the sun can efficiently reach all parts. In early forms such as the kelp, which is a large alga, the main body is a flat blade supported by a long stem. The stem is attached at one end to the rocky bottom by a root-like holdfast and the other leads to the broad blade which floats horizontally under the surface of the water. The blade then has not only the advantage of its position in relation to the sun, but its thin broad structure is effective in the absorption of the light.

The leaves of higher plants—the oval leaf of the elm, the webbed leaf of the maple, the checkered leaf of ivy, the thin blade of grass—all these are sun-catchers. They are beautifully designed for their purpose, not only in their external shape but in their internal construction and their position in relation to the whole plant. On a tree or a bush the leaves are arranged so that they do not cast too much shadow upon one another. This is especially striking in ivy where the leaves dovetail one into another at different heights, allowing many leaves to grow close together, yet each may feel the sun. In many cases the leaves even move with the sun during the course of the day by growing more on one side than the other, so that the flat of the blade remains exposed to the warm rays.

The internal anatomy of a leaf is marvelously constructed for its function (Fig. 20). On the top surface of the leaf there is a layer of thin transparent cells and below there is a layer of pencil-shaped cells all standing on end and pointing directly toward the sun. Each cell is lined with many chloroplasts, the small structures that contain the chlorophyll, and this palisade layer, as it is called, captures the light. This construction and orientation of the palisade cells is efficient in capturing light, just as an ear-horn is good at capturing sound. Below the palisade layer there is a spongy layer of cells where the movement of carbon dioxide, oxygen, and water vapor is made easy because of the large spaces between the cells. On the undersurface itself there is again a layer of covering cells, pierced with holes or stomata, and these holes can be opened or closed by two guard cells to control the amount of gas exchange and water vapor loss to the outside environment.

In a small single-celled plant all the processes take place together in one location, but in a large oak tree photosynthesis takes place only at the tips of the branches, yet the roots deep underground must grow and spread in proportion to the branches. Obviously there must be some method for conducting food to the various parts. This not only involves the transport of sugars but also proteins and vitamins and other protoplasmic substances which are syn-

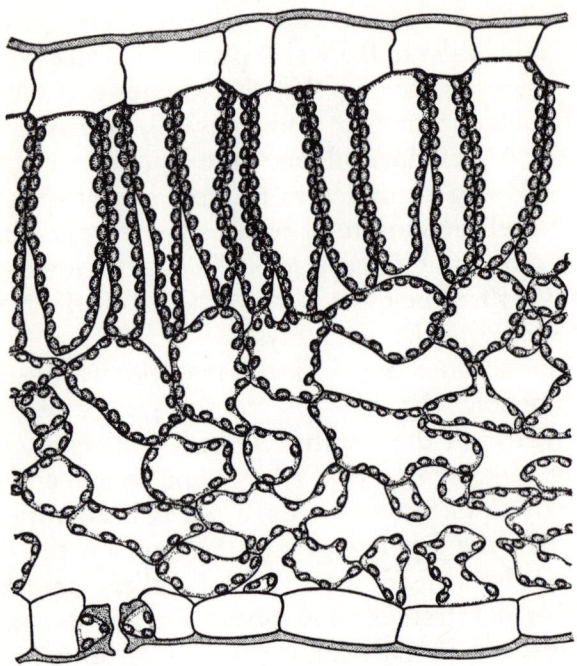

FIG. 20. A cross-section of a portion of a leaf showing the upper palisade layer and the lower spongy layer. (From W. H. Brown)

thesized in the leaves. For this water and salts are required, substances which are taken in by the roots and must be taken from the roots to the other parts. The transportation of dissolved substances in plants takes place in the vascular bundles, the bundles which are seen in the leaves as veins and in the stem as the hard parts. In a thick woody stem such as the trunk of an oak tree the whole interior wood is really a vascular conducting system, as well as a thin and more delicate layer of living cells (the bast) lying just under the bark. In all cases the conducting elements are cells, often, as in the case of wood, dead cells, which have become hollow and pipe-like to carry water with its dissolved food or salts. Many experiments have been performed to find if any particular pathways are taken by particular substances and it is now known that the water and salts absorbed by the roots go up through the internal wood, while the sugar manufactured in the leaves by photosynthesis comes down the outside bast. This latter phenomenon is especially easy to demonstrate, for if a tree is ringed so that the bast is severed in a circle around the trunk, a bulge of growth will appear above the ring because the carbohydrate is caught in its downward movement and the presence of this high-energy food in that region causes a local excess in growth. Very often the roots are involved in the storage of the food produced by the leaves, like an underground hoard, and this is the reason why carrots and beets and radishes are so plump and nutritious.

Photosynthesis and chemosynthesis are by no means the only ways in which plants feed. The great group of colorless fungi, for instance, absorbs food directly from the environment. A mold will send its delicate thread-like filaments into decaying wood or rotten meat or a host of other potential food sources and sop up the food directly. They may even parasitize a living plant or animal; and in all cases they do not synthesize their energy, they simply grow in or about the energy source. There are even cases of higher plants such as the dodder or mistletoe that rely in great part on the substance they can drain from the vascular system of the tree they parasitize, even though they also indulge simultaneously in some photosynthesis.

Most remarkable of all are the predacious forms which actually capture and devour live animals. In the fungi there are numerous examples where the fine filaments can capture small round worms or rotifer worms or amoebae. Some species of these ferocious fungi have a loop that is really a snare and if a worm should squeeze into it the loop will tighten and the fungal filament will then penetrate and grow into the dying worms until finally all its flesh has been absorbed (Fig. 21). Other species have small adhesive pegs that appear especially desirable to worms, but once some unfortunate worm has surrounded this lethal knob with its mouth, it becomes stuck and the knob grows out into the body of the worm, drinking in its substance. Still other species have spores that are eaten by the worm or the amoeba, and once inside it suddenly germinates and the sprouts soon devour the innards of the unsuspecting prey.

There are predacious plants among the higher forms too, most of them thriving on a diet of insects. The tropical pitcher plant has a leaf in the form of a saxophone and, attracted by the foul-smelling odors, insects wander in the open end only to find that they are trapped in the watery soup of digestive enzymes at the base, and any escape is prevented by a vicious array of downward-pointed spike-like hairs (Fig. 22). The fly's substance after it has been digested is absorbed directly by the plant and is distributed by its vascular system. Sundews, which we find in our temperate zones, are small plants that look like a pin cushion, the head of each pin having a drop of viscous honey on it (Fig. 23). This apparently harmless, attractive plant entices insects which become trapped by the sticky substance, and then the pin-like hairs move and hold the fly as digestive juices perform their destruction. Perhaps the most dramatic of all is Venus's-flytrap which, like the iron maiden, has a hinge and large spikes (Fig. 24). As a fly lands on the open leaf it snaps shut like a clam with extraordinary rapidity and the fly can be heard buzzing in a frenzy within its cage. There are further examples, but in each case the plant has forsaken any total dependence on photosynthesis and taken on an essentially animal mode of nutrition, for even the digestive enzymes involved are similar to those of

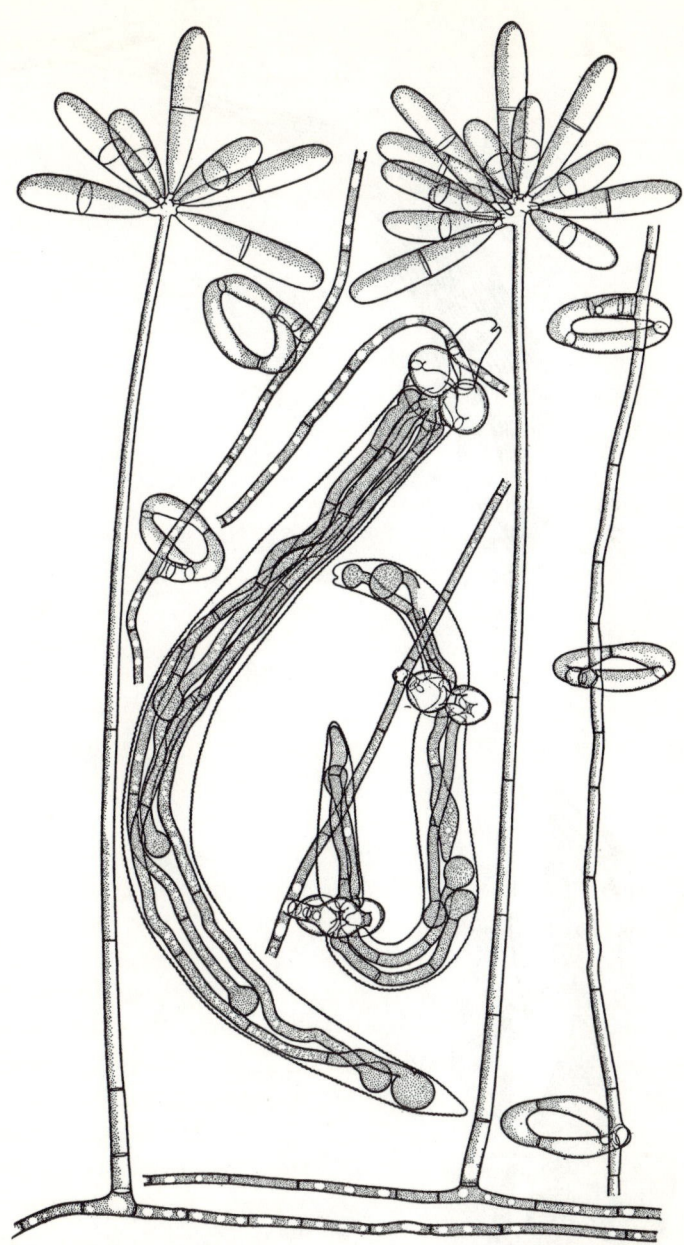

FIG. 21. A predacious fungus showing the spore-bearing bodies, the loop-like snares, and in the center one can see trapped round worms infested with fungal filaments. (From C. Drechsler)

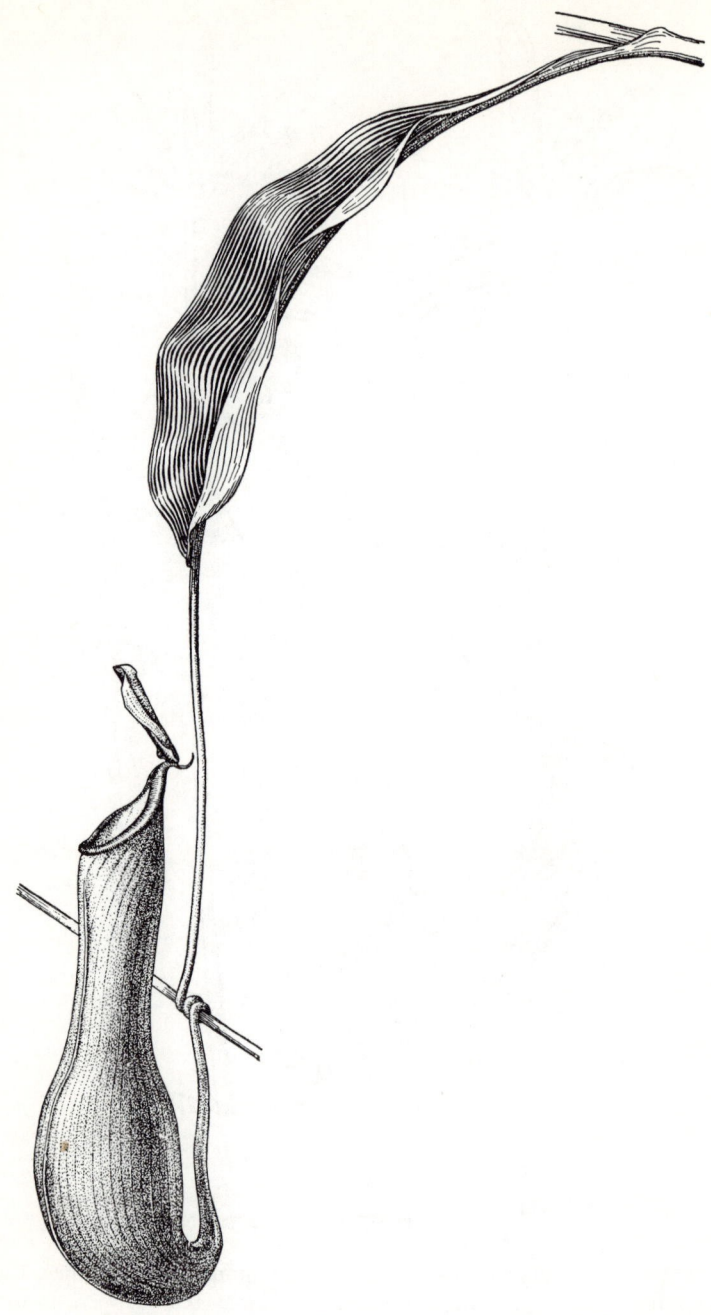

Fig. 22. The leaf of a pitcher plant. (From W. H. Brown)

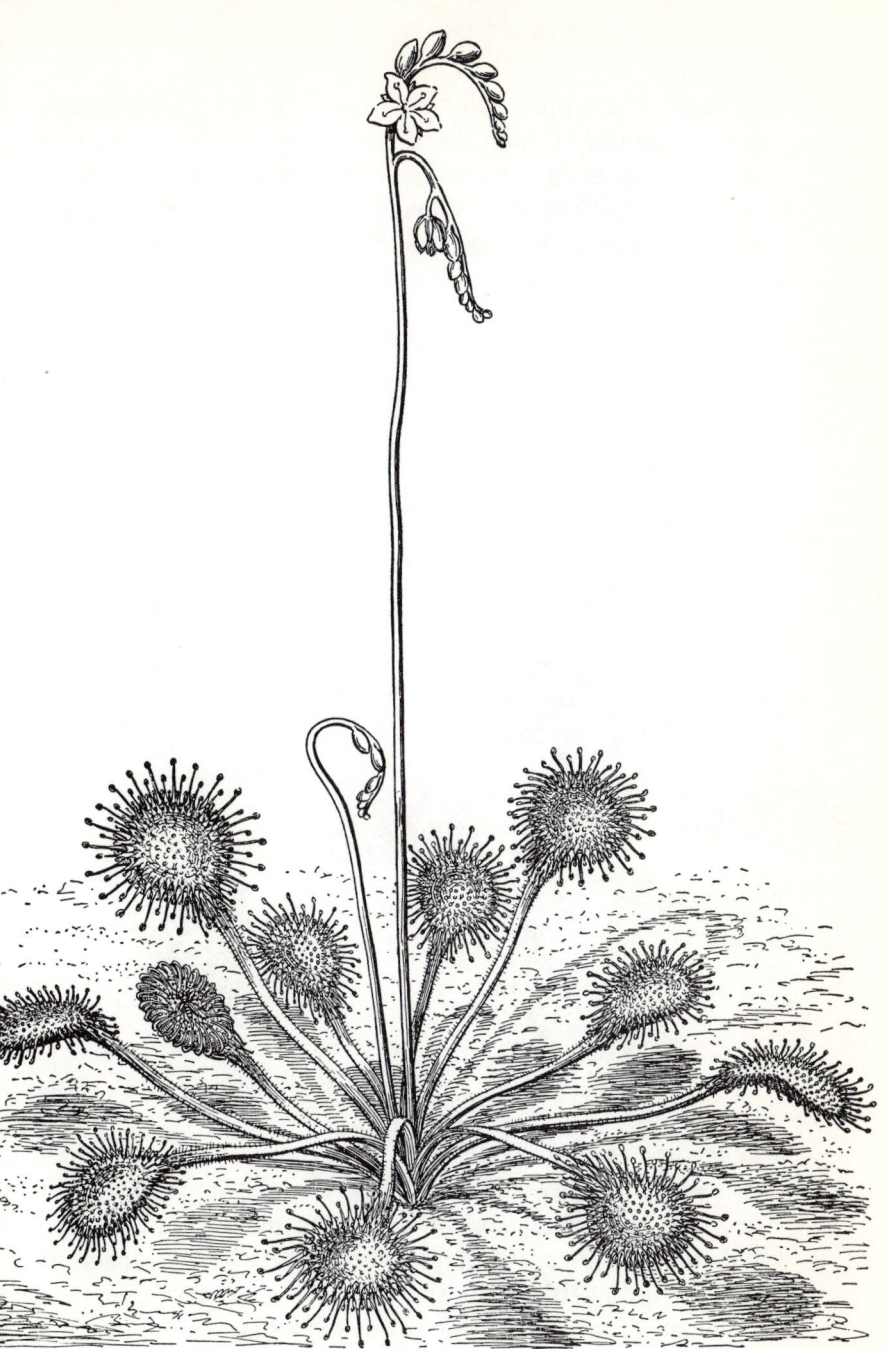

Fig. 23. The sundew, a carnivorous plant. (From W. H. Brown)

animals. These predacious plants are curious anomalies and are important in that, since they feed like animals, they show the essential sameness of all living forms. Energy is needed for all life, and it makes no difference whether the energy comes in by chemosynthesis, by photosynthesis, or

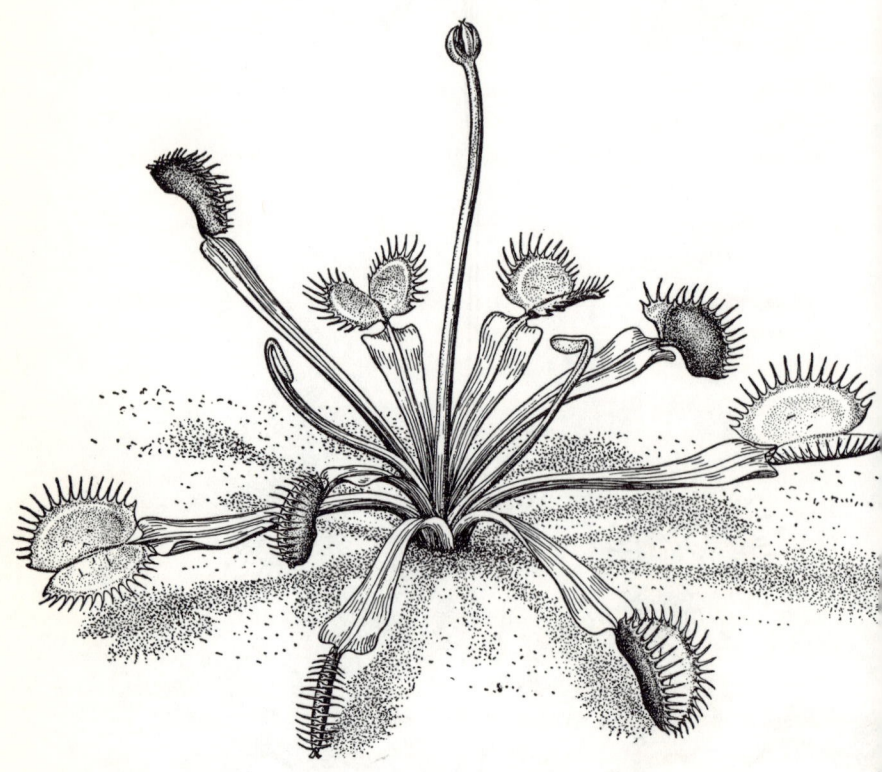

FIG. 24. Venus's-flytrap. (From W. H. Brown.)

by feeding on dissolved substances or on live animals. However each of these types of energy-intake imposes certain restrictions on the organism and often involves specific specialized structures; a chemosynthetic organism, for instance, is unlikely to achieve the size of a tree, or a photosynthetic organism of large size cannot have the shape of a sphere.

15

REPRODUCTION AND DEVELOPMENT
IN PLANTS

WE ARE concerned here with both the reproduction of cells in the growth and development of an individual plant and also the propagation of one plant from another. Among higher plants this latter kind of reproduction is largely a sexual process, and sexuality, as we have already seen, has profound advantages for the machinery of natural selection. Darwin pointed out there must be a constant source of variation for selection to take place, and this variation is greatly encouraged in sexual reproduction. We all know how human beings, for instance, vary in many ways, in physical appearance and in mental traits, and that the offspring of parents will show differences among themselves as well as differences from the parents. Each individual has some genes from each parent and these genes can combine in different ways to give different effects, producing a virtually endless source of variation, one that is completely denied in any simple asexual type of development. Therefore sexuality is a phenomenon that not only encourages natural selection, but because of this it is in turn "selected for" in evolution and is found fairly universally throughout the animal and plant kingdoms.

The most efficient kind of sexuality is the one in which the whole unit or group, such as a multicellular plant, can mate and exist in the form of opposite sexes. In howling monkeys, for instance, only the individual monkeys have this advantage and therefore the evolution and variation of the clan is bound to be a slow and perhaps an overly democratic process, since the variation of the whole clan is dependent on the minor variations among all the individuals. This may be the reason why mammal societies have never reached the complexity and the integration of insect societies, for among the insects there is only one queen (and sometimes a king), and therefore the unit of sexuality is the same for the individual as for the whole colony. From this archaic monarchy the species reaps advantages in selection on both the level of the individual and the society, perhaps explaining the extraordinary success of social insects for millions of years.

The variety of shapes and construction of plants in the plant kingdom is tremendous, and equally great is the variety in the structures associated with sexual reproduction. It would be impossible here to give any exhaustive description, but a few representative examples will be touched upon briefly. Among many algae and aquatic fungi the sex cells, that is the sperm and the egg, are motile. Often the sex cells of both sexes are identical in appearance and the only evidence that they are of opposite sex is that they fuse. In a particularly interesting and startling series of studies, the German biologist F. Moewus produced some evidence that the difference in the sex of the mating cells of a particular strain of the small unicellular green alga Chlamydomonas is caused by one chemical substance that can exist in two forms. This substance, which is related to carotene (the pigment which gives carrots their yellow color), can exist with its molecules in a right-handed or left-handed arrangement, just as one might have a right-hand or left-hand glove. In the presence of the right-handed form of the substance the cells assume one sex; in the presence of the left-handed form they assume the opposite sex. Here is what appears to be a molecular basis of a sex difference, but how the substance is responsible for this difference, and how this case applies to higher forms is unfortunately not known.

In other cases both sex cells are motile, but one is larger than the other; and in still others, such as Volvox, only the male or sperm cells are motile. In terrestrial fungi (for example the various bread molds) and again independently in the higher plants, both sexes have lost their motility and the sex cells in some way grow together and fuse.

In a flowering plant the pollen consists of male sex cells and the egg lies deep in an ovary at the base of the central pistil. The yellow, powdery pollen frequently must be carried by insects, although in many plants it depends upon the wind. An insect, seeking nectar, will rub against the male anther and become powdered with pollen, and then in turn it will rub past a sticky stigma, the knob that rises above the ovary. The pollen then germinates in this favorable environment and sends out a tube that penetrates into the flesh of the stigma, down the bottle-like neck, until it grows right into the ovary. As it pushes through and opens into the egg itself, the nuclei of the pollen tube flow down and there is a fusion of the gene groups of the male and the female. This fertilization is a complex process, and there are a series of nuclear divisions and fusions, but the net result is one cell with the chromosomes of the two parents, and this is surrounded with other cells which amount to a yolk, cells which will give nutriment to the embryo in its early development. This embryo and its surrounding food becomes the fruit, or the seed, and it slowly enlarges until it becomes ripe; in the case of the bean, for instance, it grows until the bean itself is formed. The growth of the seed involves primarily a growth of the embryo, and the original cell divides in a series of orderly divisions, enlarging at the same time and gaining its food from the surrounding yolk. It soon forms a donkey-eared structure in which the two ears are the seed leaves or cotyledons, and between these ears lies the future body of the plant: a minute rudiment which will develop into the stem, and a slightly larger posterior rudiment which will develop into the root. When a seed such as a bean, pea, peanut, or walnut is mature, the body of the seed is the two cotyledons, which are easily separated in the peanut, and between these lies the small beginning of the plant. The cotyledons are fat with food, and the young plant depends on this energy source for its food until it can expand its leaves and photosynthesize some of its own.

The seed, enclosed in its hard impermeable capsule, can be stored for many years; it is a natural stopping place in the development of the plant and one that serves the useful function of preserving the delicate embryo through the rigors of winter, ready to get an early start when the gentle spring weather comes. Then warmth will quicken its respiration and water will seep in so that all the stored food in the cotyledons will be channeled to the cells of the young plant where the synthesis of new protoplasm and the cutting off of new cells will occur with startling rapidity. The embryonic root soon lengthens out into a true root and buries itself deep into the soil, and the plumule soon stands erect as the shoot and starts the process of giving off leaves and branches. The cotyledons, now shrivelled and emaciated bags, drop off as their substance and their use has been exhausted. The plant is formed, and it has only to photosynthesize and grow, its roots penetrating deeper into the soil and its shoots reaching higher into the air (Fig. 25).

Surprisingly enough a large plant never stops growing, and it is believed that even the tallest giant sequoias continue to become taller, their only limitation being size alone, for as they grow larger they become more vulnerable to the ravages of wind and weather.

On a large plant (let us use an oak tree as an example) growth occurs at specific places. These growth zones are found in three principal regions: at the tips of the roots and the shoots, and just below the surface of the trunk. This latter zone, called the cambium, is responsible for the constant increase in the girth of the tree. It is an easy matter to demonstrate in a young shoot or root tip that growth or elongation takes place at the tip only. It is possible to

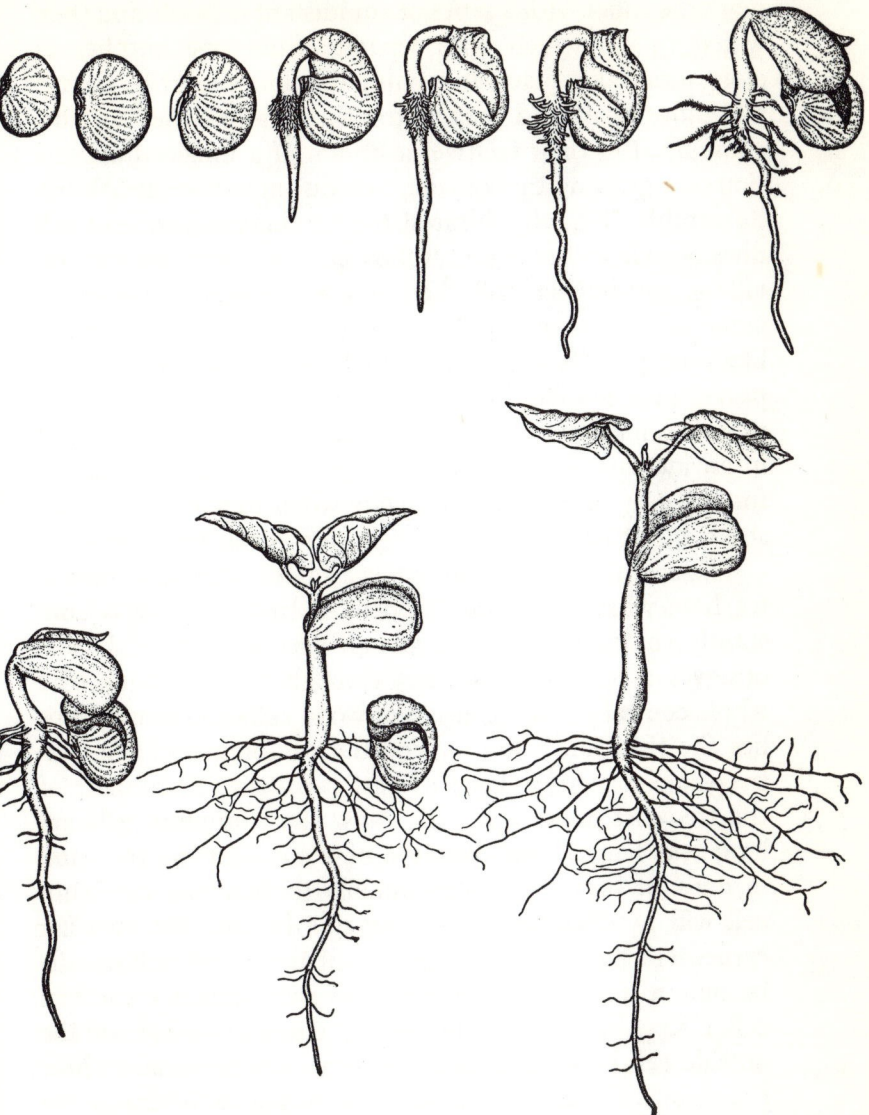

FIG. 25. Successive stages in the germination of a lima bean. The cotyledons are lifted up into the air to furnish food for the young plant and they drop off as soon as their supply is exhausted. (From W. H. Brown)

mark the shoot with a series of equidistant ink dots and then watch to see which ones separate from one another as elongation takes place. Invariably the dots closest to the tip will move apart and then as the tip advances they will become fixed in their relative positions. If a longitudinal section of a growing tip is made, two distinct zones are clearly discernible (Fig. 26). Nearest the tip there is a zone of cell division where the short, stubby cells are seen actively dividing, and further back there is a zone where the cells become elongated in the direction of the axis of the shoot, like long pencils. So the growth here has two separate aspects: first a division, followed by an elongation. Both processes involve the synthesis of new protoplasm and result in an increase in size, but the sequence of events is fixed, for it is easy for the cells to divide when they are small and short, but it becomes increasingly difficult as they elongate.

The activity of the cambium is a slightly different matter, for here what essentially happens is that the tree is constantly coated with thin layers all over its surface. If one observes a lengthwise section through a thick trunk, the whole central portion consists of wood cells and around this lies the thin cambium; outside the cambium lies the relatively inconspicuous layer of bast cells, that fleshy part of the tree just under the bark. Now the cambium cells are long cells lying in the direction of the axis of the tree, and in division they split longitudinally so that one cambium cell will produce two cells lying side by side like two flat carpenter's pencils. The inner one of these cells will usually become a wood cell, and the other will remain a cambial cell. Often, however, the inside cell remains cambial and the outside cell becomes a bast cell. It is easy to visualize how the wood will become thicker and thicker each year as the cells are plastered about the central column like so many coats of paint, but it is impossible at first glance to understand why the bast does not thicken like wood. The reason is partly that few cambial cells become bast cells and partly

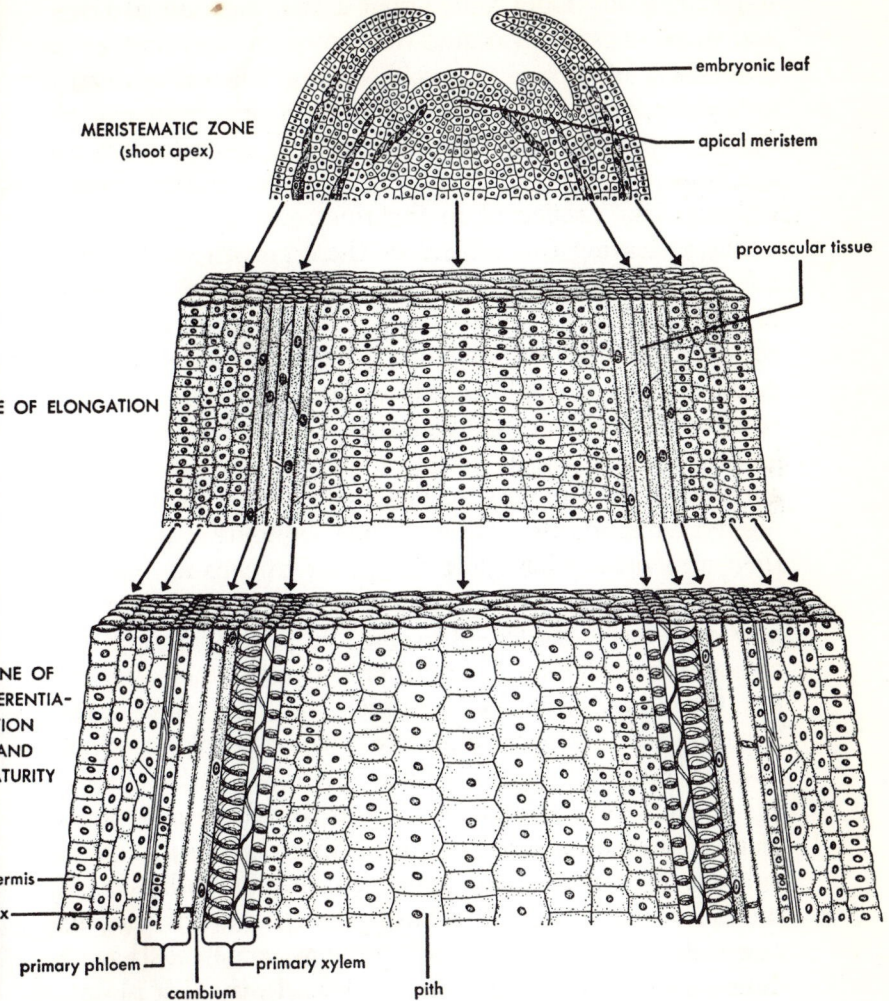

FIG. 26. The zones of growth in different portions of a stem. (From C. L. Wilson, drawing by H. T. Croasdale)

that as the bast is increased it is also destroyed, mainly by the pressure of the expanding trunk simply crushing the cells, and therefore the bulk of the tree is mainly the dead wood, surrounded by the thin living layers which circumscribe it. Every child knows that a tree has annual rings, and these rings exist because the cambium needs energy to do its work of making new cells, energy which it can easily obtain during the warm summer when its leaves are capturing fuel from the sun. In winter when the leaves have fallen the progress is slow or nil, and this difference in rate of cambial activity is reflected in the rings.

Before leaving the subject of these growth zones some reference should be made to the great advantage this type of growth affords a plant of any size. An oak tree achieves considerable size and weight, in fact it achieves such dimensions that it would utterly collapse if it were not constructed of hard and rigid materials. Cellulose is well suited for this purpose and it becomes hardened even further by another tougher material called lignin which becomes deposited alongside the cellulose. Such building materials by their very rigidity have lost the power to expand, but this is of no importance if the growth is regional, for then the new material is added onto the old, not requiring any change in the old at all. So all land plants which stick up into the air and need solid materials, and even some of the larger aquatic algae, have growth zones, and plants which have growth in all parts are invariably lowly and small.

Linked with the activity of growth zones in a larger plant is the notion of division of labor. In the first place only specific regions grow, and this in itself is a division of the labor of growing. Also during the process of forming new cells, for example in the cambium, some cells have a different fate and function from others. In the first place a cambium cell may produce either a wood or a bast cell. If it is to be a wood cell then it may develop thick walls and turn into a fiber whose major function is support, or it may develop a large hollow cavity in the center and become a conducting cell which passes water and salts up the tree from the roots. Likewise if it is to be a bast cell it may become a fiber or a conducting cell, in this case one which conducts the dissolved food sugars down toward the root. But in a large tree the division of labor, which is always elaborated in the process of development, goes even farther, for from the beginning of its embryology it shows differences in its parts; it has root primordia and shoot primordia, and each develops with its specific structures and its specific function in the overall plant. One of the important characters, then, of any multicellular organism is not only that different parts have different structures, but that these arise and are all elaborated from one germ. They are inherited differences contained in the germ, and during the process of cell division and cell growth which unrolls in a strict and orderly fashion, these differences emerge so that in the integrated community of cells the labor is effectively divided.

16

COORDINATION IN PLANTS

Most plants are extremely slow in their movements, for their movements are not equivalent to muscle movements in animals, but rather to the movements caused by growth. Because of this fact the subject of coordination in plants closely links with the problem of development of plants. If a sunflower moves as the sun moves, or if a growing shoot of a pea seedling stretches up into the air while its root buries downward into the ground, these plants are moving in their growth and the very fact that the sun is followed or the root and shoot are affected by gravity shows that these growth movements are highly coordinated.

The first evidence of how control is effected came from an interesting experiment of Boysen-Jensen who in 1913 studied the ability of shoots of oat seedlings to bend toward light. If a seedling was kept intact, its shoot showed a marked curvature toward the light; but if the tip was lopped off the shoot was quite straight, that is, unaffected by the rays of the light (Fig. 27). Therefore the tip of the shoot appeared to be vitally involved in this process of light orientation. Then he took a plant and severed the tip but reconnected it to the main part of the stem by gluing it on with a thick glob of cocoa butter. The cells in the tip were not in direct contact with the cells of the lower stem, but separated by a few millimeters of cocoa butter. When this shoot was put into the light it curved toward the source of illumination and the curvature extended below the cocoa butter junction. In other words not only does the tip initiate curvature but it can send this information through a mass of non-living jelly. From this came the hypothesis, later to be completely verified, that the tip produces a growth-stimulating substance, now called auxin, which is affected by light and which can diffuse through cocoa butter. We see that this growth coordination in plants is effected by chemical messengers, by hormones, and now let us examine the details more closely.

LIGHT

Fig. 27. A tracing of a photograph made by Boysen-Jensen showing the first experiment suggesting hormone control of oriented growth in plants. Five wheat seedlings were decapitated and in the three left-hand plants the tips have been cemented back on with a thick layer of cocoa butter. With illumination from the left, the plants with their tips replaced show curvature at the base, while the two control plants on the right show none.

The greatest advance in the study of plant hormones came from the work of the Dutch plant physiologist F. W. Went, now at the California Institute of Technology. He established a test for determining not only the presence of the growth hormone auxin but also the quantities of hormone present. This "Avena-test" involves the young seedlings of the oat (or Avena) which are placed in small racks in a row. On each plant the tip of the seed leaf is cut off and the central stem is pulled up so there is a high central shaft and a ledge around this shaft, the cut surface of the seed leaf (Fig. 28). Now a small agar jelly block can be placed on this ledge, and let us assume that this agar block contains some auxin, perhaps because for an hour or so some auxin-producing tips had been sitting on it and their auxin diffused into the agar. When, in turn, this block is sitting on the ledge, its auxin will diffuse into the cut surface of the seed leaf. But because of the central shaft the agar block can touch only one side of the plant and therefore the auxin will only descend one side of the seed leaf, and since auxin stimulates growth, this side of the seed leaf will grow more than the opposite side causing

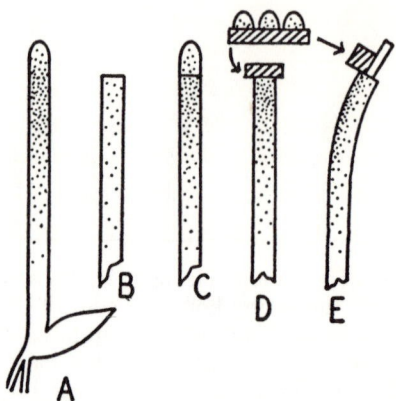

FIG. 28. Diagrams illustrating the "Avena-test." The growth rate at different regions is indicated by the density of the stippling. A, a normal oat seedling; B, a decapitated plant showing a greatly decreased growth rate; C, decapitated plant but tip replaced and the growth rate is only slightly below normal; D, decapitated plant with agar block on which the tips stood for two hours; E, decapitated plant with agar block containing auxin placed on one side (only the tissue below the block has an increased growth rate therefore causing a curvature). (From F. W. Went)

a curvature of the whole shoot. It is an easy matter then to take a shadow photograph of the shoot, measure the degree of curvature with a protractor, and the amount of curvature will bear a definite relation to the amount of auxin present.

With this wonderful tool it has been possible to make detailed analyses of the activities of auxin in higher plants. One of the first problems that was attacked was the chemical nature of auxin. This has a long and involved history with a curious twist, for it was long known that indole acetic acid, a simple substance which could be synthesized from coal tar products, was active as a growth-promoting substance and because of its specificity and availability it was used in many tests. Then recently it was shown that this substance, which was always assumed to be quite artificial, is a naturally occurring auxin actually present in plants.

It has been shown by the use of the Avena-test, and also other similar tests that have since been developed, that the principal site of production of auxin is in the tips of the plants, in the shoot tip (with its young leaf primordia) and in the root tips. It is also produced by the older leaves to some extent, and furthermore it is produced in the fruits, but the apical growing zones are the principal sources.

The activities of this chemical messenger are numerous and perplexingly varied. Perhaps the most important is the effect already mentioned: that of the stimulation of the growth in shoots. The actual chemical process of how the substance increases the growth is still not completely clear, but it is believed that the indole acetic acid enters into the cells and becomes part of the chemical machinery that manufactures new protoplasm.

I was careful to say that the growth-stimulating action was on shoots, for oddly enough the situation is quite reversed in roots. There auxin inhibits growth instead of stimulating it; there is some major difference in the tissue in its response to indole acetic acid. As a consequence many situations in the root are quite the reverse of those of the shoot; if in a shoot the tip and therefore the source of auxin is removed, the shoot virtually ceases to grow while if the auxin-producing root-tip is removed, the root, released from its inhibitor, will give a sudden spurt of growing activity.

The coordinating action of the bending of plants with respect to the sun or any light source is now well understood in terms of the activities of auxin. It was again Went who did a pioneer experiment on the shoot tips of oat seedlings, which he stuck on the end of a razor blade. The light and the blade were placed so that one side of the tip was illuminated and the other in the dark, and below each side, flush with the cut surface, he placed a small block of agar. Then after an hour or so the block of agar from the light side and the block from the dark side were tested by the Avena-test for their auxin content, and the dark-side block was found to contain a great deal more, primarily because the light actually destroys some of the auxin. If the dark side of a shoot has more auxin it will grow more than the light side which is actually deficient in auxin, in this way causing the shoot to grow or bend toward the light.

The very same argument can be applied to the root with one obvious difference. Again the dark side will have more auxin but here auxin inhibits growth so the light side will grow more and the root will bend from the light which is, of course, exactly what a root is supposed to do.

A plant is also affected by gravity, and here we find a puzzle which is not yet completely clear. There is no doubt that the lower side of a tip (let us assume that it has been placed horizontally) has more auxin, and this again can be demonstrated by the Avena-test. But how the small auxin molecules are affected by gravity is not understood. Nevertheless, the fact that auxin accumulates at the lower surface means, in the shoot, that that surface will grow relatively faster and therefore the shoot will turn and tend to grow straight upwards, away from the center of the earth. Again the response to the high auxin on the lower side is exactly reversed in the root and it will tend to burrow downward. The plant physiologists of the nineteenth century showed that in a uniformly lighted chamber, if a plant was constantly rotated so that gravity would act evenly in all directions on the plant, the perfectly opposite orientation of the root and the shoot was lost. Now we know that this is related to the distribution of the auxin near the tips.

Another interesting effect of the auxin is that of apical

dominance. The tip of the branch of a tree grows very actively. In the crotch above each leaf there are small axillary buds which are potential growth points, but they stand in abeyance as long as the growth zone in the branch tip is active. But if the apical bud is removed, then one, or in some cases two or more of the axillary buds will begin to swell and become active apical buds. This process is again associated with auxin, for if the active apical bud is removed and auxin is applied to the tip in its place, then the axillary buds remain dormant. For some unknown reason the presence of indole acetic acid itself has the power to do this, and once the substance is gone, the latent abilities of the buds are unchained. This dominance of the apex is not held to the same degree in all plants, and this difference of degree can alter the shape in ways easy to predict, for the greater the dominance, the greater the suppression of lateral branches. A conifer (a Christmas tree) which rises like a spire has a large amount, while a scrubby bush possesses little apical dominance. Gardeners have for centuries known of apical dominance, for the way to make a hedge thick is to clip it frequently so that many axillary buds will grow, and even in the case of trees, "topping" is used for the same purpose, especially in France. A similar phenomenon is found in roots, and we find tap roots and great bushy scraggly roots.

There are still more effects of auxin: while the growth of roots is inhibited by auxin, the initiation of new roots from a stem cutting is stimulated by auxin. It is, therefore, common practice now, in the propagation of plants from stem cuttings, to immerse the cuttings for some hours in a solution of indole acetic acid before placing them in the damp sand or soil. Auxin is also involved in the dropping off, or abscission, of leaves of deciduous plants. The "abscission layer" at the base of the leaf stem holds tight provided auxin is passing through the stem, and this continues during all the summer months when the leaf is healthy and synthesizing auxin among its many other metabolic activities. But with the first tinges of frost in the fall which paralyze the enzymatic machinery (or mere old age will do the same thing), the auxin goes, the abscission layer tears apart, and the leaf falls. By applying auxin at the outer end of the leaf stem it is possible in the greenhouse to prevent abscission, showing again that it is in fact auxin which is responsible. Finally, I should mention that auxin is involved in the development of the fruit, and this may be shown especially well in a few plants in which applying auxin to the ovary produces growth without fertilization, fruits without seeds.

There is no doubt that auxin controls and coordinates growth, and it plays a major rôle in governing the shape of a plant. However, when the growth hormone was first discovered it was imagined to be the whole answer, but the more that was discovered of its manifold effects, the more it became obvious that auxin alone does not satisfactorily explain the coordination of growth in the whole organism, although it does clear a great vista of understanding. There is one agent, one common stimulus, but its effects depend on the part that responds; it is the specificity of the response of different parts of the plant that gives the plant its characteristic shape. An analogy might be helpful if one imagines a garden in an arid country, and this garden represents the whole organism. In it are planted a great variety of different flower seeds (which are the different parts of the organism with their specific differences). By carefully controlling where the water (which is the auxin) will fall, with the use of a garden hose, it is possible to bring forth some plants from the soil and keep others still dormant. It is even possible by overly saturating some areas to inhibit plants if they are of a species that require little water. In this way the shape of the whole garden is determined by the water, its distribution, and the types of specific flowers and their distribution, just the way the shape of an individual plant is determined by the auxin, its distribution, and the specific types of response in the different regions of the plant.

Auxin has been found not only in higher plants but in algae and fungi as well, but unfortunately in these lower groups its normal function has not been as yet made clear. There is every reason to believe that most of the growth-coordinating activities of lower plants are, as in higher ones, controlled by hormone systems. In fact even in higher plants auxin is not the only hormone, and there are other chemical substances involved in the control of growth, such as "flowering hormones" postulated for the initiation of flowering.

But in no case in plants is there any coordinating mechanism that is equivalent to the animal nervous system. For many years it was thought that the sensitive plant Mimosa had nerves, but now it is known that this is not the case. This plant, which is a common curiosity, will fold its leaves and even collapse a whole branch if it is irritated in any way. It is a very striking phenomenon, for if the tip of a branch is pinched the collapsing response will pass in a wave and the rows of leaves will successively fall like tin soldiers and then finally the whole branch will sag. The impulse itself will travel at a speed of one to fifteen millimeters a second, but the recovery process is slow, a matter of many minutes. The collapsing is mediated by a special bulb at the base of each leaf as well as at the base of the branch and the collapsing is the result of the sudden change in permeability in this sap-filled bulb: it is like pricking a taut balloon that suddenly deflates, and in deflating the attached leaf drops. This change in permeability is caused by a hormone of a still unknown chemical constitution, and this hormone flows down the branch triggering off the basal bulbs successively. But how it travels at such a rate is still not clear, and no doubt there are special structural modifications within the plant that facilitate its movement.

There are other examples of such rapid movements in plants, and we have already seen some in the case of the predacious plants—the sudden snapping of Venus's-flytrap and the curving of the tentacle of the sundew. In each of these cases there is again a rapid change in permeability in a key structure which causes the sudden movement, and this change is effected by chemical messengers.

Whatever coordination there is in plants, then, seems to be largely by hormones; and despite many attempts to show brains in oak trees and nerve chords in sensitive plants, there is no real equivalent to the animal nervous system. In a sense there is a kind of nervous system in any cell which will suddenly change its permeability, for this is exactly what happens on a nerve as an impulse passes along it, but the plants have not evolved networks of such cells like those in animals. Yet for its purpose the chemical system of communication is highly efficient for plants, and it must be remembered that it is a system that animals also employ, for hormones play a vital rôle in animal coordination, as we shall see in a later chapter.

FEEDING AND DIGESTION
IN ANIMALS

In higher plants we found that the whole shape was modified so that they could eat efficiently, in most cases modified so that they could readily capture the sun's energy; so it is not surprising to find that in animals there are also profound structural modifications specifically designed for feeding. Success in the struggle for existence is directly dependent on success in obtaining food. In plants the source of energy is endless and all the plant needs is a place in the sun. An animal, on the other hand, must find its energy by eating plants or other animals; it must directly consume protoplasm. There is a strict limit on the amount of this kind of fuel, and the competition for it is great. One consequence of this fact is that, to avoid competition, more efficient means of capturing food will be devised, and there will be a tendency to seek exotic and unwanted foods. For these reasons there are among animals many different kinds of feeding: vegetable eaters (herbivores), flesh eaters (carnivores), and those who are not particular in their eagerness to obtain energy (omnivores). And now let us examine some examples, not shying from the bizarre cases to illustrate to what extent specialization to avoid competition can produce curious feeding mechanisms.

In the single-celled protozoa there is a tremendous number of different methods of feeding. We have already seen the way Euplotes wafts its food down a mouth-like gullet and we have talked of the false feet of the amoeba that surround and engulf its prey. A moment's reflection shows that a large share of the structures associated with lower animals are for the specific purpose of catching prey, eating it, and digesting it. The very process of locomotion, of movement, is itself connected with food-catching, and therefore limbs, muscles, and other organs of locomotion are involved. There are many devices for holding the prey, such as tentacles and stinging cells of hydra which paralyze and immobilize the potential food. Then we should consider the mouths of animals with their jaws, their teeth, and finally the whole complicated digestive tract which processes the food to make it available as an energy source. Not only are there many different kinds of feeding modifications among animals, but each one is a complex network within itself.

Among the predacious animals that capture live animals, the lion, the king of the jungle, is the symbol, the prototype and in fact his proud title comes in part from the grace with which he kills other animals and in part from his noble mien and his terrifying roar. His teeth, his claws, his stealth, and his speed all work in harmony so that even the largest of antelopes will be killed and eaten. The related cat, the cheetah, an animal much smaller than a lion with a disproportionately small head and leopard spots, is especially remarkable for its speed. For short distances it runs faster than any known mammal—up to seventy miles an hour—a burst of fleetness that soon overtakes any animal. But should it miss its prey in the first few seconds it cannot go on but must stop to gain its breath; the cheetah's power lies in the velocity of its attack.

Many rapacious birds—the owls, the hawks, the falcons, and the eagles—also kill mammals, although small ones such as mice and rats and rabbits. They will sit still in a tree or soar high in the sky and then suddenly swoop down and pounce feet first on their prey, sinking their sharp curved talons into the soft fur of the victim. Again speed is important for the surprise attack, and the duck hawk is considered to be the fastest among birds, reaching a speed of sixty miles per hour and no doubt even more in its power dives. The osprey or fish hawk catches fish in the same way by plunging on a fish near the surface and holding its slippery wriggling body in its claws. And then with its curved beak it can tear the flesh apart and greedily eat its fill.

Some turtles, such as the snapping turtle, also eat fish, and an interesting example is the Florida alligator turtle which may grow to two feet in length. It merely opens its mouth, keeping it open, and in the middle of its black tongue there is a delicious looking piece of pink flesh in the shape of a curled worm. This affords too great a temptation to some wandering fish who, in entering the snare, performs the last act of his life.

Even more grotesque are the hardly believable angler fish of the sea (Fig. 29). Arising from the middle of their

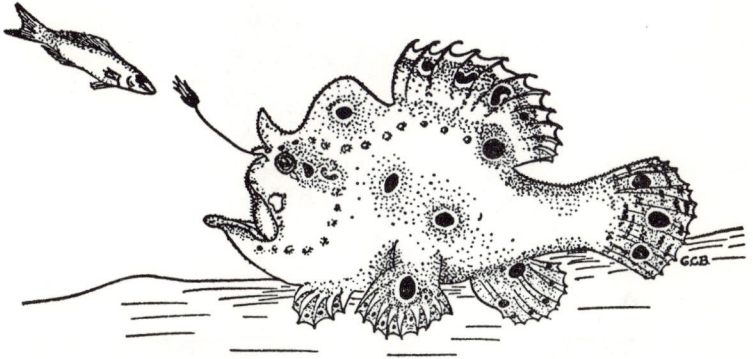

Fig. 29. The dwarf angler fish attracting its prey. (From L. P. Schultz and E. M. Stern)

back there is a long spine that curves forward and hangs down a short distance in front of the mouth, and at the end of this fishing rod there again is a piece of flesh that looks very palatable. The fish will dangle the bait in an inviting manner, and as some unwary smaller fish approaches it snaps its rod back and in a flash too sudden for the human eye to follow it engulfs the catch with its large ungainly mouth. The angler fish fishes much as the alligator turtle, the chief difference being that his equipment, like that of a weekend amateur as compared to a commercial market fisherman's, is far more elaborate.

Many animals eat insects, either the adults or their larvae. Among insects and their relatives there are many predacious forms, such as the hunting wasps or the spider with his web, or the dragon fly who, exactly like the chameleon who also eats insects, can shoot out his extendable lip with lightening rapidity and pull back the startled fly into his jaws. The ant lion, the larva of a relative of the lace-wings, is a particularly vicious form (Fig. 30). It lives in the dry

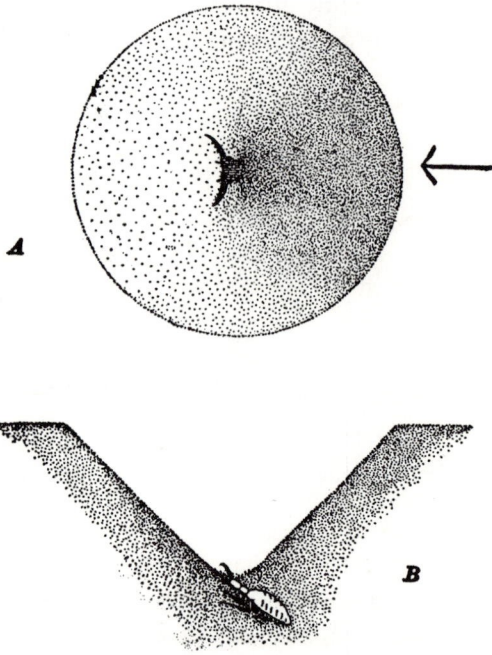

FIG. 30. The ant lion. A, the pit of the ant lion showing the position of the insect facing away from the source of light (indicated by the arrow); B, the position of the insect when lying in wait in the bottom of its pit. (From W. M. Wheeler, after Doflein)

sand and buries its body downward by throwing the sand away by a wagging thrust so that finally it is at the bottom of a small cone-shaped crater of sand, and all that can be seen of the ant lion is the sickle-shaped jaws which stick out. If an ant should pass by he will start slipping in the crater and this slipping will be encouraged by the wagging thrusts of the ant lion, making the sand slip and avalanche toward the center. The helpless ant will finally come near the jaws, and in a flash those needles will push through the hard wall of the ant into the soft flesh.

There are a number of ant eaters that are mammals; always characterized by a long snout and an especially long whip-like tongue. These beasts, who live in tropical regions where the ant hills and termitaria are large, will pull away an opening in the ant hill with their long claws and then stick their snout down a passageway. The sticky tongue will whip in and out, each time picking up hundreds, even thousands of ants on its fly-paper surface, and these are slapped into the mouth cleaning the tongue off for the next lunge. The woodpecker of our temperate zones has much the same device. Their extremely long tongue curls down into the throat region, and when they find, on the side of a tree, the opening of the larva of a wood-boring beetle or wasp, they send down their lance-like tongue which has many thistle-like barbs at the end, and these hook into the mealy body of the grub and pull it brusquely out of the hole.

Many predacious animals have specially modified mouth structures for eating small animals. The huge blue whale has a gigantic sieve across his mouth that actually keeps out any large animal, but the small shrimp can easily pass through by the thousands and millions. Clams and oysters eat minute protozoa and algae which are wafted toward a mouth by myriads of cilia lying all over the gill, and in this way they laboriously cull from the water that flows over their surface every fragment of possible energy.

Not all carnivores are predacious, and many, less attractive ones are carrion eaters. In the tropics, if an animal is dying, the blow flies will lay their eggs all over its surface, and when death comes and putrefaction begins the small larvae hatch from their eggs and eat the putrid meat. The vultures will swoop down to help them, and the filthy jackal will come to carry off his share.

There are perhaps as many, if not more, different kinds of herbivores in the animal kingdom than carnivores, but it must be confessed that they do not afford such striking examples, for all the gruesome fascination that we find for the predators is bound to go when we picture, for instance, a cow eating a pile of hay. Nevertheless the method of taking in energy is admittedly successful and in fact absolutely essential, for otherwise animals could never obtain the energy which the plants have obtained from the sun. If all animals were carnivores they would soon all eat one another up. The success of vegetable eating can be measured in a number of ways, and certainly size is one, for most of the very large animals existing today, with the exception of whales, are herbivores. The elephant, the hippopotamus, the rhinoceros, the giraffe, even the horse and cow do not touch meat. They have special teeth for grinding grass or leaves, and they have special stomachs for digesting it, because cellulose is a particularly hard molecule to break down into its component sugar molecules. This was pointed out earlier in discussing termites, and it was also pointed out that bacteria and protozoa will live in the gut of herbivores and specifically help in the breakdown of cellulose. So in some ways, with the helpful organisms inside, the digestive system of these animals is like a fermentation vat in which the end product is sugar derived from the cellulose, and this is absorbed into the body to be used for fuel.

Not only are large mammals plant eaters, but also some monkeys such as howling monkeys, many birds such as chickens, some lizards such as the giant seaweed eating lizards of the Galápagos Islands, many insects, such as the termites, the wood-boring beetles and wasps, and the fungus-growing ants. There is hardly a group that does not have vegetable feeders, each different and often highly specialized. One particular kind of vegetable feeding that is especially interesting is the eating of nectar which is found among bees, wasps, moths, and humming-birds. The interest comes for two reasons: one, the specialized feeding devices which some of the insects possess to reach the honey, as for instance the case cited by Darwin where only one species of moth had a tongue long enough to reach the nectar in one species of orchid, and the other is the flower itself which during the course of evolution has developed the nectar (and the flower as an indicator of where the nectar is) for the specific intention of attracting insects so as to facilitate pollination. Nectar is, of course, a tremendously rich source of energy, being little else than concentrated sugar, and it is because of this fortunate fact that it is an ideal food for the small humming-bird who is so wasteful of his energy as he hovers about, and with so little place in his small body to store useless weight.

Thus far, all that has been said on this huge subject of intake of energy in mammals concerns the kind of food and the method of obtaining it, and now we would do well to follow it into the body and see how it is digested. Here also the structures involved are numerous, and as already indicated in the discussion of herbivorous animals, they vary to some degree with the kind of diet, but basically the process is the same for all animals, and for that matter the same for predacious plants. The complex molecules of the protoplasmic food are broken down to small simple ones; starch and cellulose are chopped up into sugar, protein is dissected into amino acids, and fat is split into glycerol and fatty acids. Rather than confuse the reader by comparing many different digestive tracks, I shall confine myself to that of man and show the numerous structures and steps involved, and this can serve to illustrate how an individual animal prepares its food for the cells in its body, so that the cells can burn the food as fuel in their respiration (Fig. 31).

In the mouth the food is first chewed by the teeth and drenched with saliva. The saliva contains an enzyme that attaches itself to the starch molecules and begins the break-

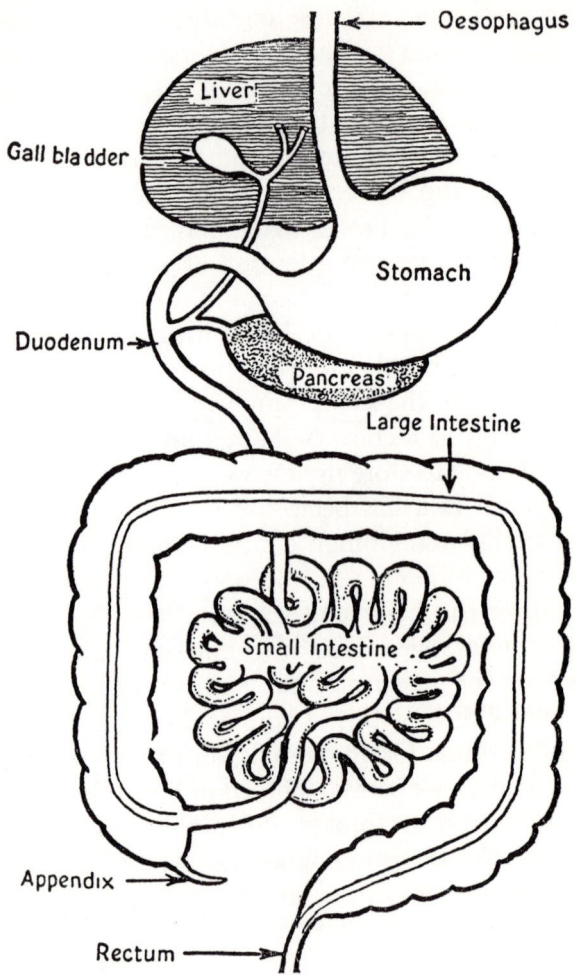

FIG. 31. A diagram of the digestive tract of a human being. (From H. G. Wells, J. Huxley, and G. P. Wells)

down of some of the starch to sugar. Then by a series of muscular contractions the food is pushed past the wind-pipe which momentarily closes during the swallowing, and once in the oesophagus that leads to the stomach the food is pushed by peristalsic squeezing down to the stomach. Peristalsis is a kind of muscle movement that occurs in tubes; the area in front of a ball of food relaxes and the area just behind contracts so that the food is forcibly pushed along, just as an orange can be pushed down a stocking by squeezing the stocking just behind the orange.

Once in the enlarged sac of the stomach, the food is churned about actively and at the same time is attacked by a battery of enzymes. The details of this process were first discovered by the early American doctor William Beaumont in some famous experiments on a wounded Indian, Alexis St. Martin. St. Martin had received a bullet wound in his stomach and it had healed badly so that the wall of his stomach grew into the hole in his skin, that is, he had a direct exterior communication to his stomach. By ingenious experiments involving feeding St. Martin different foods and removing the contents of his stomach at different times after his meal, Beaumont was able to learn many of the basic facts of the activities of the stomach, facts which have been added to in the many years of research since then. The gastric juice consists of a powerful mixture of hydrochloric acid and enzymes which break down proteins toward amino acids. Pepsin is the most important of these enzymes. It is at first surprising that this gastric juice does not digest the stomach itself, which is also made of protein; but it does not do so because the stomach is coated with a layer of mucous which protects the cells that lie below. The sensation of heartburn testifies to the corrosive action of the gastric juice, for this occurs when it slops up into the lower portion of the oesophagus and literally burns it a bit, and again in ulcers for some reason the mucous has not done its job of protecting efficiently, and the wall of the stomach itself is irritated.

When the stomach contents pass into the small intestine through a valve or sphincter, the liver and the pancreas, both major digestive glands, squirt out their digestive juices, the bile from the liver and the pancreatic juice from the pancreas. These fluids enter the small intestine at the region immediately after the stomach called the duodenum. The pancreatic juice has enzymes which continue to break down the sugars to starch, and also the enzyme trypsin which continues the breakdown of the proteins to amino acids. Furthermore it has an enzyme which turns the fats to fatty acids and glycerol. More enzymes are also produced and given off right in the small intestine, but the main function of that structure is to absorb the now available food into the blood, which in turn transports it to the cells all over the body.

The small intestine is only small in its diameter, but in length it extends a great distance (in man some thirty feet) and it is all coiled about like a dish of cooked macaroni. Its most striking feature is its inside surface, for it is covered with small teat-like projections called villi, familiar to anyone who has seen or eaten tripe. These villi serve the purpose of vastly increasing the surface area of the intestine and therefore the surface of absorption, and inside each one there are blood vessels near the surface to facilitate the transfer of the food to the blood. Both amino acids and sugars are readily soluble in water and therefore have no problem of going directly through the villi into the circulation, but the fatty acids cannot go unaided. The bile contains complex salts which surround the fatty acid molecules in a sort of layer, a cloak, and then they become soluble and in this way easily penetrate into the blood.

There is, of course, quite a bit of material that cannot be digested. This will be attacked by the bacteria that live in the intestine. They will help break it down and make some of it useful to their host. What is left passes on to the large intestine where most of the water is removed, for the short, fat, large intestine is principally a site of water absorption. The last stage in the journey is the rectum where the faeces (mainly bacteria and undigested remains) pause before being eliminated.

In our description of the energy intake of animals we have the potential energy as far as the blood; it must still reach the cells and combine with oxygen there in the process of respiration. So hand in hand with feeding there comes the problem of the circulation of the blood and the intake of oxygen through breathing. These matters will be discussed in the next chapters, for they are necessary to complete the picture of the using of energy by multicellular animals.

BREATHING IN ANIMALS

Breathing in animals might more correctly be called gas exchange, for we are concerned here with the movement of oxygen into the cells and the movement of carbon dioxide out of the cells. It must be remembered that when the living motor is running and the cells are burning fuel in their cellular respiration, then oxygen is needed to burn the sugars, fats, and amino acids to produce energy, carbon dioxide, water, and sometimes other waste products.

The problem of gas exchange is intimately bound up with the problem of the size of the animal. In small aquatic animals, such as the protozoan Euplotes, it is an easy matter for the oxygen to diffuse to all parts as fast as it is needed, for no part of the interior of the animal is more than a fraction of a millimeter away from the surface. In larger, thicker forms unaided diffusion would be quite insufficient, and if for instance a fish of any size were nothing but a homogeneous mass of cells inside, it would be quite impossible for enough oxygen to reach the interior cells and they would soon die of suffocation. The escape of the carbon dioxide is not quite so critical for it diffuses much faster than oxygen and can move more readily.

Not only will there not be enough oxygen in the middle of our hypothetical fish because the oxygen unaided cannot move fast enough through the cells, but also because the surface of a large fish becomes disproportionately small. Imagine two spheres, one twice the diameter of the other; the large sphere has eight times as much volume but only four times as much surface. Now the oxygen penetrates into an animal through its surface, and carbon dioxide escapes through the surface, therefore the surface limits the amount of gas exchange. On the other hand the requirement of cell respiration is more nearly proportional to the total volume of protoplasm.

The problem of size thus imposes difficulties on the gas exchange mechanism, and this emergency is dealt with in two ways: a regulation of the speed of penetration and the extent of the surface of exchange. The speed of transport is helped by a circulatory or blood system, a matter that will be discussed in the next chapter, and the gas exchange surface becomes specially modified, primarily by increasing its area, so that it may keep pace with the increase in the total volume of the organism.

I should like to mention parenthetically that at least one group of organisms, the jellyfish, has solved the problem in another way. Some species may reach a great size and their bell may be more than a yard in diameter, as the readers of Sherlock Holmes will know for the "Lion's Mane" is just such a form. They have no circulatory system nor any specialized gas exchange surface, yet their volume is considerable. The point is that they consist, as their name indicates, largely of jelly, and the living protoplasmic cells merely cover the surface, near enough to the oxygen so that with no difficulty they can obtain it directly. The jelly is inert, non-respiring material which makes up the bulk, and in this respect jellyfish resemble large trees where again the respiring tissue is near the surface and the bulk of the tree is dead, inert wood.

Even though oxygen is a gas, and in normal air there is about 21 per cent oxygen, it also dissolves in water and water may contain o.6 per cent oxygen. Thus all aquatic animals obtain their oxygen directly from the water. The structure involved in gas exchange is called a gill, and everyone is familiar with the gills of fishes, but many other aquatic animals among the invertebrates also have gills of different types of construction. Basically a gill is a highly ramified structure so that there is a great surface area, and it is delicate and thin so that the oxygen may easily penetrate. Usually the circulatory fluid or blood lies near the gill surface, again to facilitate the exchange, and this is seen strikingly in the gills of fishes for their conspicuous red color is nothing else than the color of the blood which can be seen directly through the thin surface. Because the gills must be delicate to function properly they are often protected, as in fish by armored side plates.

The gills of various animals are by no means evolved one from another but arose numerous times in evolution and give a beautiful example of the phenomenon called convergence. The need for a gas exchange surface is always present, and this need has been met in different ways by different animals in the course of evolution. To illustrate the point let us take some examples; in the clam or the oyster there are gills which run a half circumference about the body near the edge of the shell, like the fringe on an old-fashioned lampshade (Fig. 32); in the lobster or shrimp there are feather-like branches that stick out from its lower surface; in the sea cucumber, a form related to starfish and

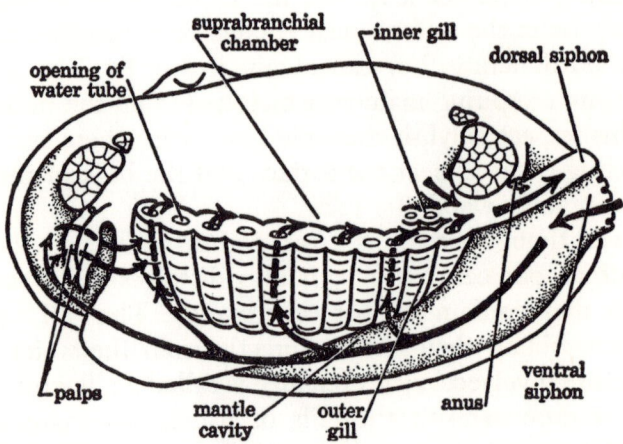

Fig. 32. Diagram illustrating the circulation of water in the gills of a fresh-water mussel. (From F. A. Brown, Jr., and M. E. Pierce, after Wolcott)

sea urchins, there is a gill attached inside the anus and gas exchange takes place at its posterior end. None of these gills are directly related to one another, but just as animal societies arose independently many times, so did the gills of animals. The only similarity that these structures have is the activity they perform.

Many gills have a further property that aids the exchange of gas. If a gill is quiet and the water about it does not move, soon that water will be depleted of its oxygen; therefore circulation of the water would be very advantageous to supply fresh water laden with oxygen. The lowly sponges show this principle admirably, for inside a sponge there are thousands of flagellated cells, and with their whips they move the water. It comes in from pores all over the surface and shoots out one large hole at the top of the sponge (Fig. 33). The English zoologist G. P. Bidder made a fascinating study of the hydraulics of sponges some thirty years ago. He showed that sponges living in still water follow sound engineering principles, that is, the outgoing hole and the water pressure are such that the waste, oxygen-depleted water is shot far enough away so that the sponge does not immediately use it again. But in sponges that live in tidal currents no such mechanism has evolved, and the sponge just spreads itself like a fan so that the moving water passes through it, and the sponge need not bother to move the water.

In higher animals with true gills there are various types of ventilation. In clams the water is moved, as in sponges, by small cilia all over the surface of the gills. Crustaceans such as shrimp or barnacles wave their gills through the water. In fish that live in still water the gills flap open and closed, a familiar sight, and the purpose is to bring fresh water to the gas exchange surface. It is interesting that in many fish that are fast swimmers, such as the tuna, this ability to ventilate is lost. These fish simply open their mouths and keep swimming, for as in the sponges that live in tidal currents, the aerated water will constantly flow by.

A gill is admirably suited for an underwater existence, but in the conquest of land that occurred a number of times during evolution the gill is, of all the structures of the body, the least well adapted for existence in the air. The main

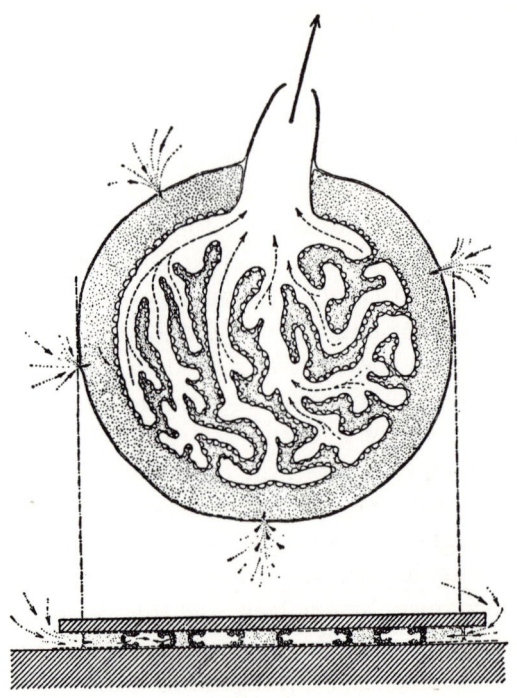

FIG. 33. A diagram of a sponge grown (as indicated below) between two slips of glass. In this flattened condition it is readily possible to see the canals and the direction of the water currents all eventually leading out one large opening. (From W. E. Ankel and H. Eigenbrodt)

reason for this is that in order to capture oxygen a gill must be thin and wet (the latter so the oxygen can dissolve and penetrate in solution) and it is an extremely difficult matter to keep such a delicate fleshy structure from shrivelling up in the presence of dry air.

Despite these difficulties some animals do have air gills.

There are a few species of snails and a few crustaceans that do this but either by keeping the gills fairly covered or living in a moist environment such as wet sand, they manage to preserve their gills. The best known case is that of the peculiar fish of the mangrove swamps called Periophthalmus (Fig. 34). It has strange popeyes on stalks and uses its

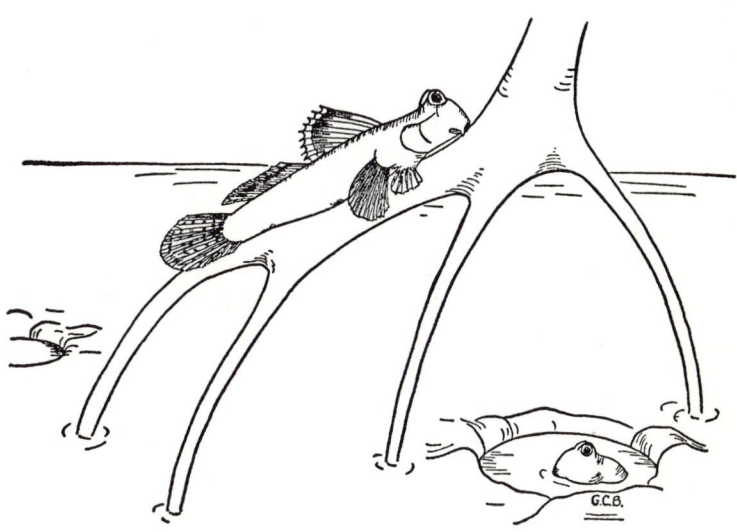

Fig. 34. Periophthalmus, the gilled fish that climbs out on mangrove stumps. (From L. P. Schultz and E. M. Stern)

fins for feet, in this way waddling out on the mangrove stumps and remaining in the air for many hours. But this takes place in the damp tropics of Siam where the relative humidity must be close to one hundred per cent at all times, and therefore there is little danger of the gills drying up.

Now in the fishes there have been numerous attempts to produce a lung rather than a gill. The difference is that a gill is exposed directly to the air but a lung is confined in a cavity, and the great advantage here is that by virtue of being a cavity it can retain moisture permanently. We, of course, have lungs, and our exhaled breath is saturated with water vapor, an easy matter to achieve in such an enclosed space. The most successful example in fish is where an out-pocketing is formed in the upper part of the gut, near the back of the throat. Such is the case of the African lungfish which lives in fresh water in places that often dry up, so the lungfish must survive long periods of drought. Most modern fish have lost the ability to breathe with this lung and use this air sac or "swim bladder," which lies just under the spine in the abdominal cavity, as a method of controlling buoyancy.

There are many examples of lungs quite independently evolved among invertebrates; and certain snails, the pulmonate snails, have a quite specialized lung. The lungs of insects, called tracheae, are radically different in that instead of being sacs they are small branching tubes that, like blood vessels, have finer and finer ramifications so that these gas canals come into contact with practically all the cells, greatly reducing the function of the blood, for here the gases can be taken directly in and out of the deep tissues. There are paired openings on each side of the body for each segment of the insect, and in this way all the parts are reached. There is a special problem of ventilation because it is difficult for the gas to flow of its own accord up and down these small tubes. For this reason body-wriggling in insects and the beating of the wings (even the slow pulsing of the wings of a butterfly at rest) tend to keep the air moving in the tubes and help bring in fresh air.

The lungs of vertebrates higher than fishes, the amphibians, the birds, and the reptiles are paired structures which lie in the chest region and are connected by a common tube to the back of the throat. In many of the amphibians, which of course are gilled in their larval form, such as the tadpole of the frog, the lung is a simple affair not more than a single sac. It is often asked how is it that a mature frog, which has no gills, can stay under the water practically indefinitely. The answer is that its skin is thin and its whole surface acts as a gill and takes in oxygen directly, and when it comes out of the water it uses its lungs.

Reptiles, birds, and mammals, most of which are basically independent of water completely, have a harder impermeable skin and can only breathe air with their lungs. It is interesting that as the size of the animal increases the surface area within the lung increases; the lung is divided up into many minute compartments or alveoli, in this way always improving the efficiency.

In all higher forms with lungs there is a need for ventilation, a need for pushing the air in and out of the lungs to replenish the source of oxygen and eliminate the carbon dioxide. In frogs the animal simply swallows the air and then burps it up again, a rather rudimentary and only fairly satisfactory procedure. In the reptiles and higher forms the chest cavity is sealed off and with the ribs forms a large pumping bulb. The muscles expand the chest and the air is sucked in, and by relaxing the muscles the chest contracts and the air blows out. The lungs themselves do not contract and expand; it is only that they are in a large sealed vessel that does, and since the lungs have an opening leading to the outside, they fill and empty with the movements of the chest (Fig. 35). Only in the huge elephant is the lung attached to the chest wall, and in this case the lung is physically distended and relaxed. Birds are especially fortunate in that when they do their most violent exercise, that of flying, the chest is expanded and contracted with each wing beat making it possible to combine the movements of locomotion and breathing.

There are many other aspects to the physiology of breathing in mammals which might concern us here, but for lack of space I shall use but one to serve as an illustration of the extent to which breathing is a coordinated and perfectly regulated process. The reader is well aware of the fact that if he holds his breath he soon is possessed with an overwhelming desire to breathe, and it takes considerable will power to keep from doing so; in fact this is an impossible method of committing suicide. The reason is that as soon as one stops breathing, the concentration of carbon dioxide builds up rapidly in the lungs and the blood as a

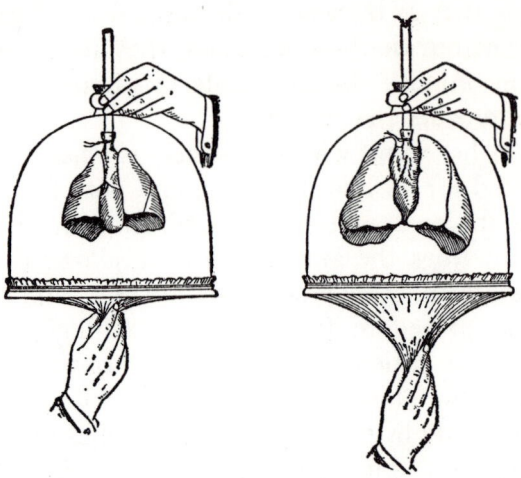

Fig. 35. Diagram showing the inflation of the lungs being accomplished by the production of a suction or negative pressure in the chamber surrounding the lungs. (From Woodruff after Tigerstedt)

result of continued cell respiration, and there is a nerve center in one of the principal blood vessels which is sensitive to carbon dioxide. That is, if the carbon dioxide is high, this nerve is stimulated and it in turn stimulates the nerves that control the muscles involved in the breathing movements. Therefore one of the most effective ways of holding one's breath for a long period of time is to force-breathe in and out hard for some time, eliminating as much carbon dioxide as possible. The abnormal lack of carbon dioxide will bring a certain tingling giddiness, and when that moment arrives one should start to hold one's breath. If a few lungfuls of pure oxygen are taken at the last moment, this gives an even further advantage and by doing this a student in Professor Edward Schneider's laboratory at Wesleyan College held his breath as long as twenty-one minutes. There are still other regulatory mechanisms so that the body manages, by an intricate series of checking systems, to maintain itself in a steady state, irrespective of what kind of a physical task it is performing. And all this delicate balancing machinery is directly related to the simple fact that all the cells of the body need oxygen to burn food and obtain energy.

CIRCULATION IN ANIMALS

TAKING energy into larger multicellular animals involves three major processes: eating and digesting the food, the breathing or gas exchange mechanism, and finally the circulation. The flowing liquids of the body are involved in the energy exchange in more than one way, for in the first place the blood transports oxygen and carbon dioxide from the lungs or the gills to the cells of the body. Also the blood carries the food from the villi of the small intestine to the body cells, and after cell respiration has occurred the waste products of the cells are carried off to be eliminated through the kidney. Not only that, but the hormones of the body are carried by the blood, and one part of the body can pass information to another part by mailing a sort of chemical letter that is carried by the postal service that is the blood stream.

In the various groups of invertebrate animals there are different kinds of blood systems. In some of the lower worms there is no more than a fluid enclosed in the body and by the wriggling of the body the fluid moves about. In others the fluid is partly enclosed in vessels some of which have regional concentrations of muscle that contract rhythmically to give a rudimentary heartbeat that helps circulate the blood. The annelid worms, such as the earthworm, have an efficient and well-constructed circulatory system, as do the molluscs, especially the higher cephalopods (squid, cuttlefish, octopus, and nautilus), whose hearts are elaborately constructed and efficient in their operation. The need for a transporting system is always felt in larger organisms, but the method by which this is achieved varies greatly and has evolved independently a number of times. Because of the great variety it would be impossible here to give any comparative anatomy of circulatory systems; and even in the vertebrates, in the fish through the mammals, there is an extraordinary evolutionary change. All the blood systems of vertebrates stem from one ancestral type, but in the course of evolution the modifications are numerous and profound, tending generally to give an increasingly efficient type of circulation.

One change that has occurred is in the way the blood is aerated in the lungs or gills. In fish there is a two-chambered heart: the first is an auricle which slowly fills with the blood from the body between beats, and then it contracts and pushes the blood into the other chamber, the highly muscular ventricle (Fig. 36). The ventricle gives a sudden power-

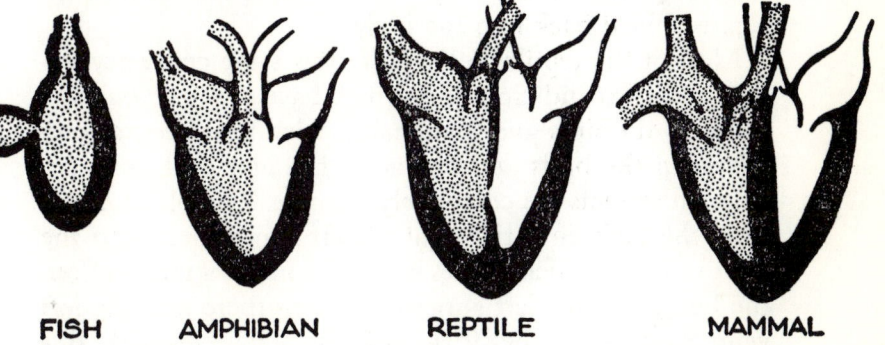

FISH   **AMPHIBIAN**   **REPTILE**   **MAMMAL**

Fig. 36. A comparison of the hearts of different vertebrates. The stippling indicates the blood which is devoid of oxygen. (From L. A. Kenoyer and N. H. Goddard)

ful contraction and forces the blood through the gills and from there it goes all through the body to eventually drain back again into the auricle of the heart. In the next group, the amphibians (frogs, toads, newts, and salamanders), there is a three-chambered heart and a separate circulation in the lungs. The blood drains into the right auricle which pumps it into the one large ventricle. There are two vessels that lead from the ventricle, one to the body and one to the lungs, so that the blood does not have to go in series through the lung and then through the body as it does in fish. The advantage here is that considerable pressure is involved in pushing the blood either through the lungs or through the body and by shortening the circuit and separating it into two, the available power is greater. But the disadvantage of the amphibian circulatory system is that there is no distinct separation of the aerated and non-aerated blood, for after the blood has been aerated in the lung it returns to the other auricle (the left auricle) which shoots it directly into the ventricle again and mixes it with the used blood coming from the body. So in the ventricle of amphibians there is a mixture of the two bloods and there is bound to be a certain degree of inefficiency as it shoots some already aerated blood back into the lung and some non-aerated blood back into the body.

Most of the reptiles suffer from this same deficiency, except alligators and crocodiles which have a four-chambered heart, as do birds and mammals, including of course ourselves. In the birds and mammals the lung and body circulation systems are completely separate. The blood enters from the body into the right auricle, from there into the right ventricle which pushes it into the lungs; it returns from the lungs into the left auricle and from there into the left ventricle which pushes it through the body. Obviously this last push of the left ventricle is by far the greatest, and therefore it is the most muscular part of the heart. Our heart lies slightly to the left because of the increased size of the left ventricle.

The internal structure of the mammalian heart is especially interesting. In the first place the muscles themselves which constitute almost the entire wall are of a special kind. The very fact that they can beat many times a minute for the lifetime of a man is enough indication of their remarkable properties. Connecting the chambers there are huge valves which in themselves do not contract, but they are forced shut or open depending upon the internal hydraulic pressure. If the ventricle contracts the doors leading back to the auricle where the blood comes from slam shut, and the doors leading out to the artery open. The position and shape of the valves cause them to act as they do. There are many other interesting aspects of the heart, for instance the control of the rhythmical contraction and the control of the rate of the beat, all of which involve a careful balance of nerve and hormone coordination. If an animal runs fast he needs more oxygen, and therefore the blood must move faster and the heart must beat faster. This all happens automatically, without the volition of the individual, by means of a remarkable network of regulatory control mechanisms.

The fact that blood travels in a circle was first discovered in the seventeenth century by William Harvey. In a celebrated treatise he laid down many proofs of the fact that the blood leaves the heart through the arteries, penetrates the tissues of the body, and returns through the veins. A short time later the inventor of the microscope, Antony van Leeuwenhoek, discovered the capillaries. As the arteries approach a tissue, they break up into smaller and smaller vessels until finally throughout the tissue there is an intricate network of minute tubes or capillaries (Fig. 37). As the blood leaves, the small capillaries come together again in successive stages, finally to be collected in one large vein that leads back to the heart.

Because the capillaries are so fine and numerous, their surface area is great and therefore friction imposed on the blood is correspondingly great. The heart, especially the left ventricle, has to be very powerful in order to push the blood through the capillaries. The pressure, therefore, is greatest on the arterial side of the blood circuit, before the blood reaches the capillaries, and it is for this reason that the arteries are thicker than the veins. Arteries also must have great elasticity to yield with the rhythmical pumps of the heart, yet because they are elastic they do not dissipate the actual driving force of the blood. In hardening of the arteries, a disease that sometimes accompanies old age, the arteries lose their resilience and the pressure changes between

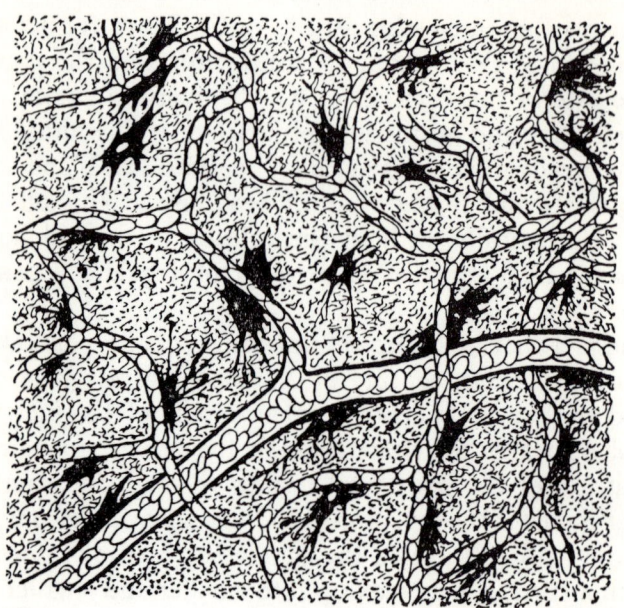

FIG. 37. A fine artery branching out into capillaries, as seen in the frog's foot. (From H. G. Wells, J. Huxley, and G. P. Wells)

one part of the heart beat and another are felt directly on the tissues, often causing serious destruction.

By the time the blood gets through to the veins the pressure has gone way down, and in fact under certain conditions the blood has a hard time getting back to the heart. Along the major veins there are valves which keep the blood from going back the wrong way, and a number of other artificial devices are needed to help, especially to bring the blood way up from the legs. In the first place, as was said before, each time one takes in a breath a suction is created in the chest cavity, and this suction helps to draw the blood up from the lower part of the body into the main vein that leads to the heart.

Another artificial aid in getting the blood back to the heart is the movement of the muscles, which inevitably squeeze the veins and push the blood upward. If you have watched or perhaps stood in a review parade at attention for a long time, the importance of these muscle movements is often made clear, for it is a common sight to see a soldier suddenly faint and collapse on the ground. The reason is that the blood has become stagnated in his legs; it cannot struggle back up to the heart without the help of the muscles, and without blood to provide oxygen for the brain, consciousness goes. The officers in charge, however, are quite right in doing what appears to be cruel—to leave the soldier in a heap on the ground—for in that position the blood will soon get back to his head and he will be alert again. A more gruesome example of the same phenomenon is death by crucifixion. If a man is tied to a cross he will die because he cannot move his legs; he will die because the blood will pool in his legs and his brain will have a continuous unabated loss of oxygen until finally the fainting spell becomes irreversible.

Thus far we have talked about blood as a liquid but have said little about its chemical properties. In lowly invertebrates its inorganic chemical constitution closely resembles that of the sea, and in higher forms there are some differences. It has been argued and with some reason, that when vertebrates first evolved, the salt constitution of the sea was slightly different than it is now, and that vertebrate blood reflects rather closely this primeval ocean. The blood also contains organic compounds; and besides the food, the waste products, and the hormones there are various serum or blood proteins which are the important part of blood plasma. But most interesting of all are the blood pigments, for these colored compounds are involved in the transport of oxygen.

If oxygen simply dissolves in the blood, it cannot reach a concentration higher than 0.6 per cent, for that is the maximum solubility of oxygen in water. Since this is inadequate for most larger animals, the problem is solved by combining the oxygen chemically with a blood pigment, and this chemical combination is of such a delicate nature that once in the tissue the blood pigment readily gives the oxygen up to the cells. In the presence of much oxygen, as in the lung or gill, the pigment avidly seeks out the oxygen, but when the external oxygen concentration is low, as in the deep tissues, the pigment lets the oxygen go.

These blood pigments are of different sorts in different animals, although in all cases they are complex proteins, associated with a metal atom. Some of the worms have a beautiful green blood in which iron is involved (chlorocruorin); the insects and relatives for the most part have a blue copper pigment (haemocyanin), while all the vertebrates and one isolated group of insect larvae have a red iron pigment (haemoglobin). Now if it is remembered that only 0.6 cubic centimeters of oxygen can dissolve in 100 cubic centimeters of liquid, it is possible to compare the combining power of the blood of different types of animals. Molluscs, for instance, can hold between 1 and 2 cubic centimeters of oxygen in 100 cubic centimeters of blood; annelid worms (e.g. the earthworm) can hold from 3 to 10; fish, amphibians, and reptiles from 5 to 15; mammals from 15 to 20; and finally diving mammals such as whales and seals that must hold their breaths for long periods of time can keep as much as 40 cubic centimeters of oxygen in 100 cubic centimeters of blood.

How is it possible for the blood of different animals to have different combining capacities with oxygen? In the first place the blood pigment itself may be more efficient chemically to perform the combination. Also, and this is a more important factor, the quantity of pigment in the blood may be greater. In the various invertebrates the pigment is simply in solution, and the more pigment the more concentrated the blood. But it is impossible to continue any such simple concentration process to a significant degree, because the pigment molecules are large proteins and as they become concentrated the blood becomes hopelessly viscous, and if pushed to an extreme it could no more circulate in the small vessels than honey, and the heart would be impossibly taxed. The vertebrates have avoided the problem by not keeping their haemoglobin in simple solution, but instead they keep the molecules packed in small containers, the so-called red blood cells. Inside each red cell there can be a tremendous number of haemoglobin molecules, and provided the cells are not too abundant, the blood will remain thin and flow easily.

Some twenty years ago an expedition headed by the British physiologist Joseph Barcroft went to the high Andes to study the adaptation of animals and man to high altitudes. They found that at fourteen thousand feet the Indians could play a violent game of soccer while the members of the expedition found it exhausting merely to watch. It is true that the Indians who are born and raised at high altitude have abnormally large chest cavities, apparently a result of growing up in an atmosphere containing little oxygen. But the principal difference was that the blood of the Indians was much thicker than that of the visitors, that is, the number of corpuscles per unit volume of blood was much greater, and this was true of the wild animals also as compared with their brothers in the lowlands. After a month or so, as the explorers became acclimatized, they also found an increase in their red blood cells, which meant that they could exercise more readily at high altitudes. The process of the adaptation of the blood to high altitude is slow, while upon returning to sea level the loss of red blood cells occurs rapidly.

Nothing has been said so far concerning the elimination of carbon dioxide. This mobile gas does not really need much assistance; it can leave practically of its own accord. There is in the red blood cell a special enzyme which speeds its separation from the blood, but with this one bit of assistance it goes from the cells to the blood and out the lungs with no difficulty.

The circulatory system affords a vital link between the exterior of the organism and the innermost cells. It carries the food, it carries the gases, it removes the wastes, it faithfully serves all the inward parts so that they can obtain and convert energy.

EXCRETION IN ANIMALS

THERE is a problem connected with the intake of energy in animals that does not really exist in plants, and that is the need for removing waste products of the energy machine, the exhaust. There are a number of reasons why plants are not bothered with a problem of waste disposal. In the first place they burn mainly carbohydrates that produce carbon dioxide and water, neither of which requires any special mechanism of elimination, and also plants are great accumulators, so should there be any undesirable waste substances they are usually just parked somewhere in the plant in the form of crystals.

Animals on the other hand eat proteins, substances which contain nitrogen, and the waste nitrogen must be eliminated even though the process of doing so represents a considerable problem. The majority of the lower aquatic forms, the protozoa, sponges, coelenterates, etc., eliminate their nitrogen in the form of ammonia gas. This substance has the great advantage of being extremely soluble, so it diffuses out readily, which is important since ammonia itself is a toxic substance and if it did not escape quickly it would soon paralyze the animal.

Perhaps because of this toxicity many higher aquatic and terrestrial animals convert the waste nitrogen into urea, which has the advantage of not being toxic and which is moderately soluble in water. We eliminate our nitrogen primarily in the form of urea, which passes out directly in the urine. Urea does impose one condition on the animal: that it must pass sufficient water so the urea can go into solution. For fish in the sea who can drink water, and even for human beings, this is no problem, but it is a serious

191

difficulty for the clothes moth larva that lives on dry wool and must therefore conserve every drop of water it contains; and it is a problem for a bird or reptile embryo that lives for a long period in an impermeable egg and has no way of taking in or getting rid of water.

For them urea would not be a useful substance, and instead they manufacture the more complex nitrogen-containing substance, uric acid. This acid has the property of being almost completely insoluble and it forms small crystals in the body. In insects such as the clothes moth larva, these crystals are simply collected in one region in what is known as a kidney of accumulation, and they are more or less permanently stored there. This provides no inconvenience to the larva since the crystals do not take up much space; it is a compact way of packaging the nitrogen. In a reptile or bird egg the crystals of uric acid merely accumulate, and there is no toxic effect on the developing embryo. Once the bird has hatched, it eliminates the uric acid directly, mixed with the faeces, although in a free-living bird this affords no particular advantage as it does to the embryo in the egg.

Now it should be mentioned that the kidneys of animals are not concerned solely with nitrogen excretion, but also with the problem of salt and water balance. In some aquatic forms the water-salt concentration inside the body is simply the same as that of the outside, but in most aquatic and all terrestrial animals the internal water-salt quantity remains nearly the same irrespective of the environment. This requires delicate mechanisms of adjustment. If the salt in the blood is too high, salt must be eliminated, or if the water content is low, water must be conserved and not recklessly poured out.

The mechanism in salt-water fishes is interesting, for the fish drinks sea water which contains far too much salt, but the salt is excreted in special glands on the gill, in this way maintaining the blood salt at a constant level, lower than the environment. In sharks, on the other hand, the concentration of dissolved substances inside is kept the same as the sea, but by a peculiar mechanism: they do not excrete all their urea but maintain it in the blood, and thus, although the substances differ inside and outside the shark, the total concentration is the same. This has been the cause of a number of financial failures in the canning industry for if shark meat, which is very palatable, is put up in cans, the urea comes out and crystallizes all over the surface of the meat, giving an unappetizing appearance, to say the least, as the housewife opens the can.

As we have already seen, water balance is maintained in the protozoa by the contractile vacuole, a sort of bilge pump that operates at such a rate that there is a constant differential between the interior and the exterior osmotic pressures. It is unlikely that this vacuole is concerned also with nitrogen elimination since the ammonia gas can so easily escape directly through the cell wall.

In higher forms there appear a variety of different kidney structures. In flat worms there are small "flame cells" that have brush-like flagella at the end of a small tube, and as the wastes are put into the tube, which leads to the exterior, the undulating flagella help push them out. In the earthworm the structures involved are different and more elaborate, as they also are in insects, and again we cannot pause to examine in detail each one of these organs of elimination that arose independently in evolution. Instead we will discuss the human kidney briefly so that the general principles of kidney function may be understood.

Our kidneys are paired bean-shaped structures about the size of a closed fist, located in front of the spine. Large blood vessels lead to and from each of these kidneys. A tube leads from each kidney down to the centrally placed bladder, the reservoir of urine, and there is a further tube from the bladder to the exterior. Normally a human being passes about one and a half quarts of urine a day, although by drinking as much fluid as possible an individual may pass as much as fifty quarts a day.

The kidney itself consists of millions of minute tubules, each of which contains a cup or a capsule at one end, and there are two regions surrounded by cells, the proximal region near the capsule and the distal region near the collecting tubes which finally lead to the bladder (Fig. 38).

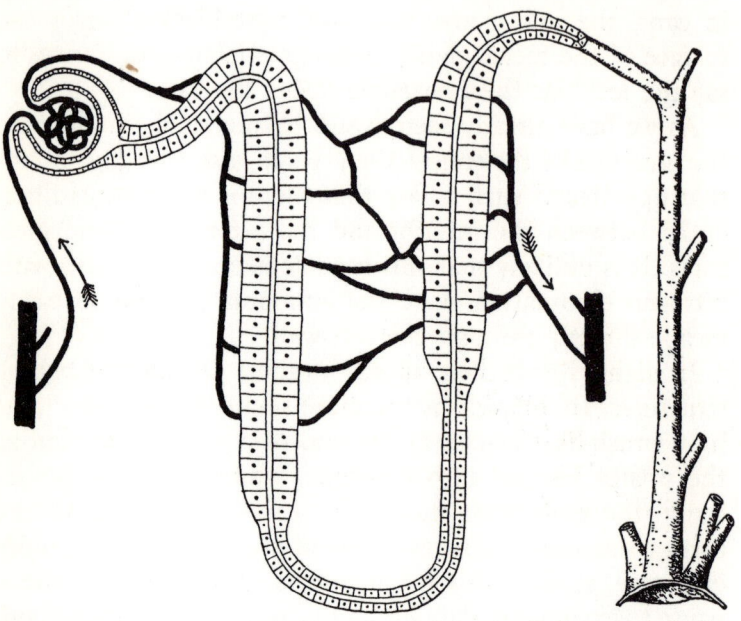

FIG. 38. A diagram of the kidney tubule of a human being showing the blood supply, the capsule, the proximal and distal portion of the tubule, and finally the collecting tubules which lead to the bladder.

There is a small blood vessel which comes to the capsule and there spins about like a tangle of wool (the glomerulus), leading finally to an intimate contact with the cells of the proximal tubule and then with the cells of the distal tubule, returning finally into the main circulation of the body.

When the blood arrives at the capsule it is laden with urea and other kinds of minor nitrogenous wastes, perhaps excess salts, and certainly excess water. All the components of the blood except the cells and the large proteins such as serum proteins filter into the cup and begin their voyage down the tubule. When they reach the first proximal portion the cells there begin to pull out and put back into the blood vessels those substances that the body wants. It is a case of Indian giving; for instance sugar may have gone through the capsule, for it is a small molecule, yet the body obviously wants all the sugar, so this is reabsorbed from the crude urine and put back into the blood. Also not all the water is allowed to go, and some is brought back. Next the distal tubule does the same thing, but with the salts that are needed and must be retained. So by the time the urine has passed these two customs houses, nothing is left behind but the undesirable wastes, the urea, the extra salts, and the extra water. It is an elaborate and delicate mechanism perfectly regulated so that the blood maintains a normal healthy composition. Without this exhaust mechanism, this blood cleansing, the energy turnover machinery of the body would soon suffocate itself and the living activities would be irrevocably stopped.

21

DEVELOPMENT AND REPRODUCTION
IN ANIMALS

BASICALLY the problem of reproduction in animals differs hardly at all from that of plants. There are in the lower forms numerous means of asexual reproduction involving budding and fission, but sexual reproduction is the predominant means, and in higher animals the sole means. The advantage of sexuality in producing variations for evolution is as true for animals as it is for plants.

The physical means of uniting the male and the female sex cells is extremely varied in animals, for here the organism itself possesses motility, and in the classic although by no means unique situation, the male can pursue the female. In aquatic forms there is a whole gamut of methods for uniting the sex cells. In some hydroids, for instance, there is a simple simultaneous squirting of the sperm and the eggs into the water, and the timing is governed by the light. Hydractinia, whose colonies we have already examined, is such a form, and after almost exactly one hour in light preceded by a longer period of total obscurity, both the sperm and the eggs are shed; it is a beautiful and dramatic sight to watch in the laboratory as the round green eggs float free in the water and the myriads of white sperm squirt out like wisps of smoke.

In higher aquatic forms such as the fishes there is also a simultaneous shedding of egg and sperm, but here the mechanism of controlling the simultaneity is often far more elaborate, and instead of depending solely on external factors such as light, internal psychological factors are involved. The Dutch animal psychologist N. Tinbergen has shown this especially well in the case of a small fresh-water fish, the three-spined stickleback, where there is a courtship ritual—a series of stimulatory movements on the part of one member of the pair, and corresponding responsive movements of the other which in turn stimulates the first member—and by this complicated succession of mutual stimuli and responses, this pantomime sex dance culminates in the synchronous shedding of egg and sperm.

Another good example is the squid or cuttlefish, highly developed molluscs that not only have an involved courtship, as Tinbergen has again shown, but also special structural modifications for copulation that are amazingly elaborate. The male has a special copulatory arm within which the sperm are enclosed in a complicated sort of package. After the courtship and display, the male clasps the female and places his special tentacle in the mantle of the female. The arm then breaks off at the base, leaving both the sperm sac and the special copulatory arm in the female. Later the sperm escape from the sac as the female lays the clusters of elliptical-shaped gelatinous eggs.

As a rule in aquatic forms, and this is true of the cases cited, the egg and the sperm unite free in the water, and the swimming ability of the sperm permits this to take place. With the conquering of land, animals were faced with the same problems as plants, and without water some modification of the system of fertilization was necessary. Plants, as will be remembered, solve the problem in a number of ways: in some forms such as terrestrial fungi the sex cells just grow together and fuse, while in shrubs and trees the light or sticky sperm are carried by the wind or by insects to the female part of the flower. The motility of animals has permitted a quite different solution of the problem, and independently a number of groups—the annelid worms such as the earthworm, the insects, and then again the reptile, bird, mammal series—have developed internal fertilization. By copulation the male introduces the sperm directly into the female. Amphibians represent an interesting transitional case; they need water for external fertilization, but the male frog, for example, will clasp the back of the female, and as the female lays the eggs he squirts the sperm directly on them as they emerge. But even in cases where fertilization is internal there is an important psychological element, for the sexes must be brought together and stimulated to unite. An individual must therefore recognize a member of the opposite sex, and often in animals with sexual seasons, mating can only take place when the eggs and sperm are ripe. Frequently there are courtship displays which will progressively excite the pair to unite, and these mechanisms operate only when the animals are in perfect physiological condition for such a union.

The first step in the development of the offspring is that of fertilization, the penetration of the sperm into the egg. This process is an interesting and surprisingly complex one that still is far from being completely understood. The sperm is little else than a flagellated nucleus (with half of the normal number of chromosomes). The egg, on the other hand, has a relatively large amount of cytoplasm, and in many instances, such as the bird, a huge quantity of yolk as well as a nucleus (again containing half the normal number of chromosomes). As the sperm swim about, often in thick clouds about the egg, one sperm, and under normal conditions one sperm only, will penetrate the surface of the egg, usually leaving its tail behind, stuck to the surface. A membrane then appears and lifts outward from the egg surface, in that way still further isolating the myriads of unsuccessful suitors. The male and the female nuclei are now together in a common cytoplasm, and they fuse to form one nucleus with the normal body complement of chromosomes (half from each parent). Then this nucleus divides, the chromosomes duplicate themselves, and each daughter cell has what appears to be the same constitution as the original fertilized egg. Differences do appear in the cells upon subsequent divisions, but this matter of differentiation will be discussed shortly.

Fertilization itself involves a series of interesting chemical reactions in which both the egg and the sperm participate. First the egg produces a substance which tends to agglutinate or clot the sperm, and the sperm in turn produces a substance which clots the eggs, and these two substances presumably help bring the egg and the sperm together and keep them stuck to one another. The sperm also produces a substance which facilitates penetration, but the egg, upon being fertilized produces a substance which helps to prevent further sperm from entering.

The details of the progressive changes from egg to adult differ in important ways in different animals and it would be quite impossible to describe all the different ways. One of the factors which governs the structural changes involved, aside from the obviously key matter of the shape of the adult, is the amount of the yolk present in the egg and the related question of the position of the yolk (Fig. 39). If there is a small amount, as there is in the eggs of many invertebrates such as the star fish or clams or numerous worms, then the yolk is evenly distributed to form a fairly homogeneous distribution in the cytoplasm. In others, such as the egg of the frog, there is so much yolk that the egg has two regions; one contains largely yolk, and this gently grades into the region containing the majority of the protoplasm and the nucleus. The consequence of this is that in division the cells do not separate quite so readily at the yolk end as they do in the other and soon the protoplasmic end is quite populated with small cells while the yolk end will have much fewer, larger cells. In fish, birds, and reptiles the amount of yolk is so great in proportion that the protoplasmic part just sits like a small drop on the surface of the yolk, and the development instead of being three-dimensional and spherical is disc-like and flat, at least in its earlier stages. Most mammals do not have so much yolk, but they also have this flat kind of early development inherited from their immediate ancestors the reptiles.

The flatter the development, the more difficult it is to follow its sequence, and therefore for the sake of simplicity we will follow the development of Amphioxus, the lancelet,

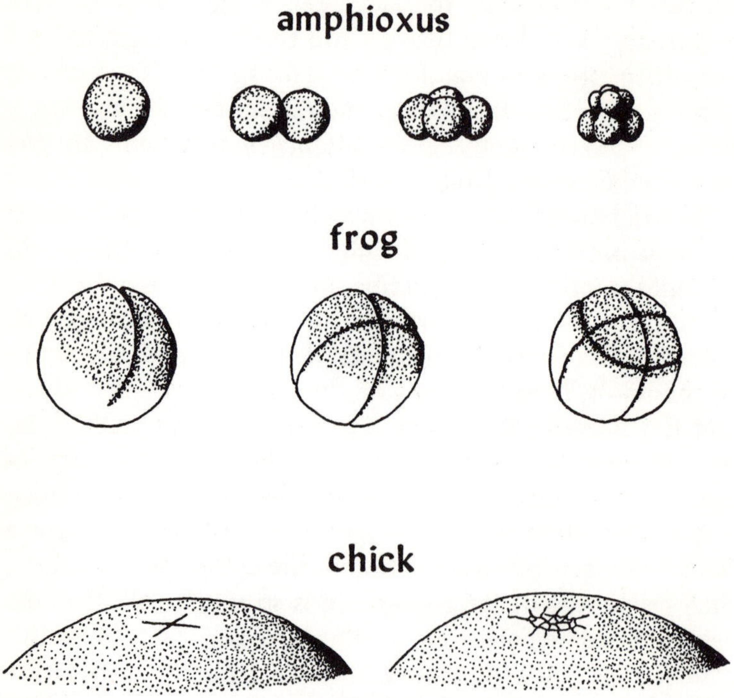

amphioxus

frog

chick

Fig. 39. A diagram showing the development of three different kinds of eggs containing different amounts of yolk. The top row shows cleavage in the eggs of Amphioxus, the middle row is that of a frog, and the bottom row is that of a chick.

a forerunner of the vertebrates, which has a simple body plan and a diagrammatically clear development that neatly demonstrates all the basic principles of the geometry of animal development (Fig. 40).

Since the yolk is fairly homogeneously distributed in the

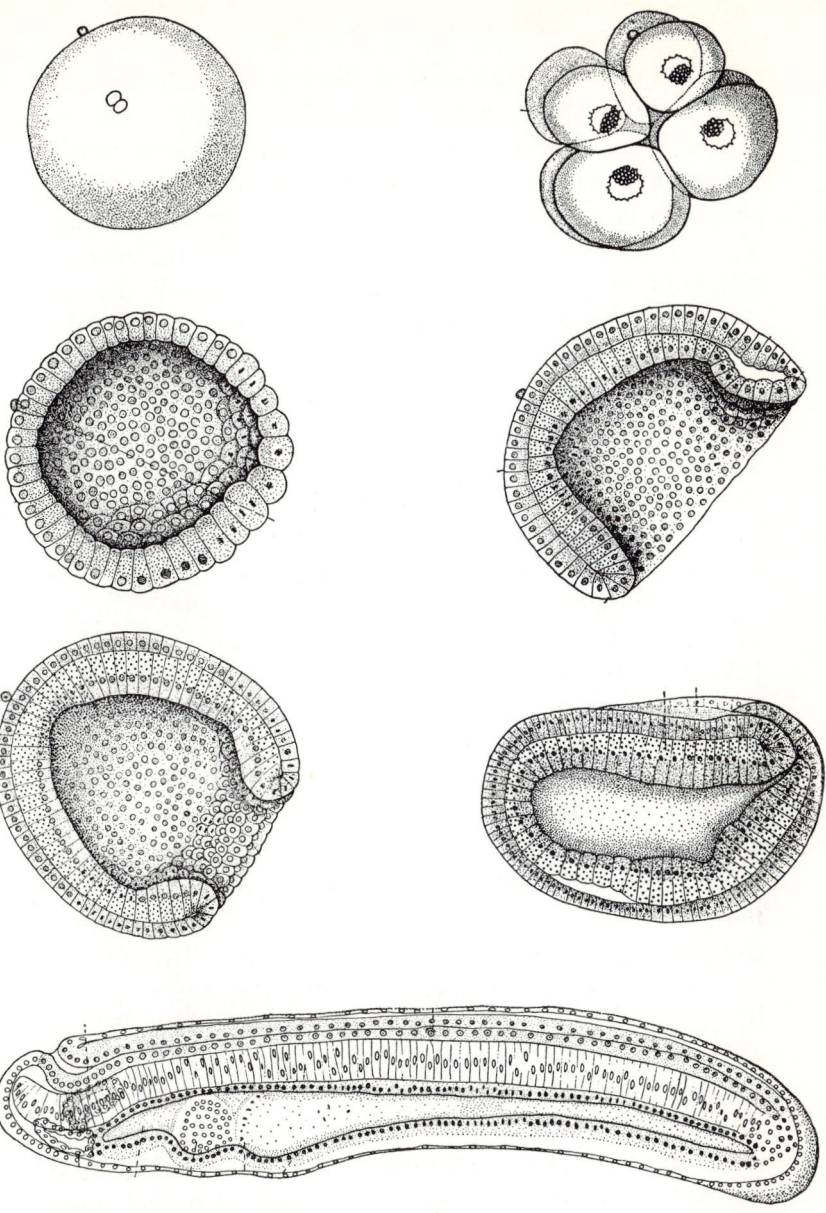

Fig. 40. A series of diagrams showing the development of Amphioxus, starting with the egg on the upper left and ending, in the bottom figure, with the major parts of the body blocked out. (Drawings by E. G. Conklin)

fertilized egg of Amphioxus, it divides evenly in two, and then in four, eight, sixteen. Even when there are only sixteen cells it is clear that a cavity is beginning to form at the center, and as division continues, the cells become progressively smaller as the cavity becomes progressively larger. This so-called blastula stage is similar to Volvox, a hollow sphere of cells. But it soon disappears and slowly becomes converted to a gastrula with one side pocketing inward so that the blastula cavity virtually disappears and a new gastrula cavity is formed. It is as though one took a tennis ball and pushed one side in to make a double-walled hemisphere. The inside wall of the gastrula is destined to become the endoderm (the gut and its associated structures) although a small part will pocket off to become the mesoderm (the muscle, and the supporting structures, i.e. the notochord which is the rigid primitive spine of Amphioxus). The outside wall of the gastrula is destined to become the ectoderm (the skin and the nervous system). Now let us follow these structures in more detail.

The gastrula grows and in doing so assumes an increasingly elongated shape, as the lips of the cup come closer and closer together, like the opening of a purse pulled shut with draw strings. Once the opening of the gastrula, which is the future hind end of the animal, is nothing but a small pore, the ectoderm grows over it in a flap and begins to cover the upper back surface of the embryo. This sheet of ectoderm then forms two ridges, with a depression between them, running up and down the length of the back. Like waves the ridges grow toward one another, and the hollow between them continues to drop so that finally a tube is formed running the length of the back, and this is now the elementary spinal nerve chord. While this is taking place the endoderm forms three similar ridge-like pockets along the length of the embryo, just under the nerve chord pocket. The central one of the three becomes the stiff notochord, while the one on each side develops into muscle, and soon invades all the area between the skin and the gut. The animal is now blocked out, and all that need be done is to add and finish the details. The nerve chord sends out nerve branches and a nerve network; the notochord becomes rigid; the muscles assume their final position and form and start to transform their cells into contractile units; the gut grows further pockets which are the digestive glands, and finally an opening appears at each end of the gut to form the mouth and the anus.

The mechanism of development in animals, and particularly the mechanism of differentiation, is a problem that continues to puzzle and fascinate biologists. How is it that an organism that develops from one cell, the egg, can produce the tremendous division of labor among its cells, some cells becoming nerve cells, others muscle cells, others notochord or bone cells, and so forth? Not only are the cells differentiated to divide the labor, but so are the organs which they form, the liver, the kidney, the lungs, etc., and even the organ systems such as the blood system or the nervous system or the digestive system. All this remarkable division of labor springs from the egg, that one cell that appears misleadingly simple, and we should all very much like to know exactly how this process takes place. In many ways the same problem was seen in the formation of new ant colonies, for there in the beginning only the queen exists, and later come all the different workers with their particular jobs. In ants it is known that in some cases the difference among the workers is a difference in their nutrition while growing, but how this special feeding is controlled is a more involved problem. In certain kinds of developing eggs nutrition may also play a part; for example in frogs' eggs certain cells have more yolk food than others, but more than this is needed to explain all the detailed cell differences.

There are two possible ways in which differentiation might take place: as the cells divide specific key substances might become segregated differentially in the daughter cells (mosaic development), or possibly the cells might be potentially identical but with their fate determined by their position in the body (regulative development). There are good examples of both; for instance in the ascidians, which are not too distantly related to Amphioxus, the late E. G. Conklin of Princeton University showed some fifty years ago that even before the first division after fertilization the different parts of the egg, which have prominent color markings and therefore can be recognized and followed, are destined to become specific parts of the larva that will follow. And if, by cutting, those parts are removed early in the development, the larva will be correspondingly deficient and will lack those parts.

At about the same time the German embryologist Hans Driesch performed a similar experiment with different results. He removed part of the early embryo of a sea urchin (related to a star-fish), but the resulting larva was, although small, quite normal and possessed all its parts (Fig. 41). He pointed out that the cells had equal potentialities and that a certain cell turned into a specific type only because of its position on the whole organism. If it was on the surface it became a skin cell, but if it was on the inside layer of the gastrula it became a gut cell, and so forth.

At first it seemed that these two types of development were quite opposed and incompatible, but nowadays that is no longer felt to be the case. It is postulated that basically all eggs or germs are of the regulative type, but at some moment in their history they become rigid and inflexible and mosaic. In ascidians, as Conklin showed, this occurs early in development, but in sea urchins the mosaic character does not set in until later.

Much work has been done, especially in Germany by Hans Spemann and his school starting some forty years ago, on the development of amphibians, frogs, and salamanders; for there, as in sea urchins, development is of the regulative type for a considerable period in the early development. Spemann found during gastrulation that a peculiar process took place: the tissue just above the hole where the pushing-in occurred to form the gastrula had a special ability to

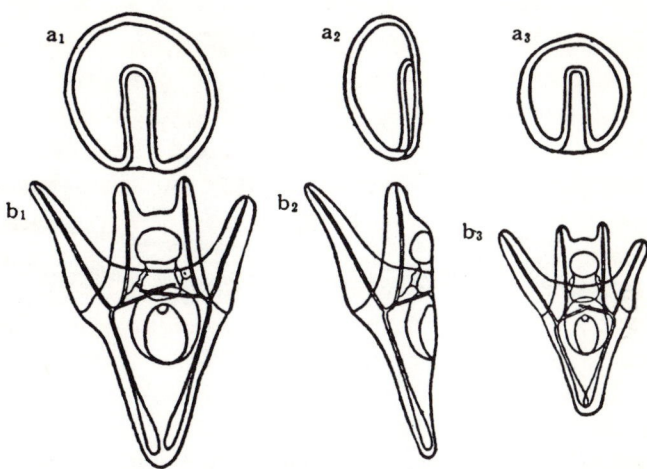

FIG. 41. Driesch's experiment with developing sea urchin larvae. a_1 and b_1 are the normal gastrula and the normal larva which result; a_2 and b_2 are what Driesch expected to observe by allowing a half embryo to develop; a_3 and b_3 show what actually occurred, namely a perfect dwarf gastrula and embryo. (From H. Driesch)

induce the tissue it touched to form the main axis of the embryo, that is the spinal chord, the nerve chord, and all the associated structures. If a piece of this special tissue was grafted onto another embryo, then that embryo would produce twins, one by its own special tissue, and another by the grafted tissue added (Fig. 42). This remarkable region Spemann called the "organizer" and for a while its properties were a complete mystery, but then one of his students, J. Holtfreter, found that he could boil the organizer tissue, thereby killing all the cells, and it would still induce a new embryo in the tissue it touched. From this it became clear that a chemical substance is produced in the organizer region which stimulates the cells to do what they are already potentially capable of doing, namely forming an embryo.

This situation, then, parallels closely that of auxin, the growth substance in plants, for in both cases there is a chemical stimulation and a beautifully coordinated re-

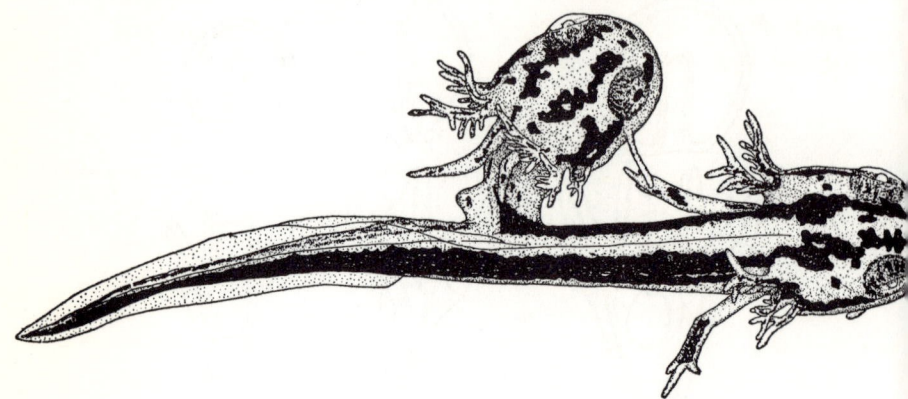

FIG. 42. A two-headed larva of a salamander. (Note: this was not actually produced by a graft but it serves to illustrate the appearance of a twin in an amphibian.) (From G. Fankhauser)

sponse. But as we saw in the discussion of plants, it is not enough to explain the form of an organism by simply saying there is a substance that stimulates and there is a precise response. For there are many other aspects: the specific location of the stimulating substance and the types of response it evokes in different regions. There is a whole complex network of interdependent stimuli and responses, and our research has merely scratched the surface. The constancy of proportions of living organisms from generation to generation is a marvelous fact which we do not completely understand. We saw the same problem in animal societies, where for instance termite colonies kept only one royal pair and would produce another only if the original pair were killed. There, some of the mechanisms, through the interesting work of Lüscher and others, are beginning to be understood and may even guide us in our research upon the development of multicellular organisms. But to me one of the most striking aspects is that despite the complexity of termite societies, they are in truth really extremely simple developing systems compared to the frog embryo. There may be many parallels between the two, but animal embryos are likely to hold out longer against the complete analysis of the biologist.

22

COORDINATION IN ANIMALS

IN AN ANIMAL society there are many different methods by which information is passed: gestures and vision may be involved, or calls and hearing, or odors and smell. It is true that in each case this information is recorded by the sense organs of the individuals, but the information itself is passed by different physical methods, sound, light, and chemical diffusion. Inside an animal there are two possible ways by which information is passed: that is by hormones, chemical substances which diffuse into the blood and are carried to different parts, or by nerves which carry impulses directly like telegraph wires.

In its broad aspects, the hormone system of animals is similar to that of plants, although it is a more complex affair, with more numerous hormones and more involved interrelations. The glands that produce the hormones are variously called glands of internal secretion, ductless glands, or endocrine glands; and as all these names imply, they do not secrete their hormones into a duct which leads away, as do the digestive glands such as the pancreas with its pancreatic duct, but instead they secrete their hormones directly into the small capillaries that perfuse the gland. As a matter of fact the pancreas is also an interesting example of a ductless gland because, while most of its cells produce the digestive enzymes that flow into the duodenum, there are small groups of cells, the islets of Langerhans, that give off hormones to the blood. The principal hormone of these islets is insulin, a hormone which controls the amount of sugar in the blood. The presence of insulin allows the excess sugar to become stored in the form of glycogen or animal starch. When a human being has deficient islet cells he suffers from the disease diabetes, but he can live a normal existence if he artificially injects insulin into his blood stream.

Another important gland is the adrenal gland, found in pairs just above the kidneys. The outside layer of the adrenals produces a number of substances, some of which are directly involved in maintaining the salt balance of the individual, and a malfunction of this part of the gland causes Addison's disease, which, like diabetes, is fatal unless hormone injections are continuously taken. The central core of the adrenals produces the hormone adrenalin, which has manifold effects, all of which are related to the problem of preparing the individual for greater stress or activity: the heart beats faster, the amount of sugar available for immediate use is increased, the respiration increases. Sudden anger or sudden fear will stimulate the secretion of adrenalin to prepare the body for an emergency. It is a common experience—you can feel a flush and the thumping of your heart against your ribs.

The thyroid gland is a fairly large butterfly-shaped gland that lies just behind the position normally occupied by a bow tie. The gland produces a hormone called thyroxin, and thyroxin is important in governing the rate of cellular combustion all over the body. If there is a deficiency in thyroxin, the metabolism will be low and the individual will be slow and lazy; but if the gland is over-active and the thyroxin excessive, he will suffer from too much energy and never be able to keep still, never be able to stay on any one conversational topic for more than a minute; in extreme cases his eyes will be quite popped giving a restless, desperate appearance. If the gland is deficient from birth, the child will not grow normally and will be dwarfed with peculiar crimped features, but these so-called cretin dwarfs can be cured if caught early enough, by thyroxin injections. Another interesting aspect of the thyroid gland is the problem of goiter, a disease in which the gland swells to grotesque proportions. In each molecule of thyroxin there are atoms of iodine, and therefore to manufacture thyroxin, iodine is needed. In most localities there is fortunately enough iodine in the soil so that in eating a normal diet a human being receives a sufficient quantity, but in certain regions, the goiter belts, there is a deficiency of iodine, and unless it is artificially added to the diet people living there frequently have goiters. There are such regions in the center of the United States and also in Switzerland, and it is for that reason that most of the salt one buys at the corner grocery has been "iodized," that is, iodine has been artificially added to it.

It must be made clear that this list of hormones is oversimplified and abridged to provide a knowledge of the kind of chemical coordination that exists rather than a complete picture with all the details. For instance, the minute parathyroid glands which lie near the thyroid and which are involved specifically in the control of calcium in the blood will be no more than mentioned, and instead we will devote a bit more space to the minute pituitary gland, for it is the central headquarters of the whole endocrine system of mammals.

In human beings the pituitary gland, which is about the size of a pea, lies at the base of the brain (in the back of the roof of the mouth) and is actually attached to the brain. However, despite its small size, the number of known hormones it produces is considerable, and there are probably some that have simply not been unmasked as yet. One of the important hormones it produces is the growth hormone, which stimulates the normal growth of an individual and in this way closely resembles the plant hormone auxin. If the pituitary gland is deficient in producing this hormone, the child will remain a dwarf, although his features and general proportions will remain quite perfect. If, on the other hand, because of some tumorous growth on the pituitary the gland oversecretes the growth hormone, the individual will become a giant. There are cases recorded of boys in their teens reaching over eight feet in height, but this exaggerated growth weakens the health in general and these unfortunate individuals rarely live long. Normally at about twenty the long bones of the body become sealed off at each end and cannot increase in length any more; they lose their growth zones once they reach a certain age. But if a tumor should now strike the pituitary and a sudden increase in growth hormone occur, then the body as a whole cannot increase, but there are certain cartilaginous regions in the body such as the joints in the fingers and in the jaw region and nose region that still can grow, and the result will be a peculiar distortion of these regions called acromegaly.

The principal functions of the pituitary involve interrelations or coordination of all the other endocrine glands. For instance the pituitary gland produces a hormone which stimulates the thyroid to produce thyroxin. It also has a similar relation to the adrenals and to the pancreas. This is usually not a one way channel but a reciprocal action, that is, the outlying glands send hormones to the pituitary which in turn sends its messages back. This is seen particularly well in the menstrual cycle of women, for during the course of each cycle there are two different hormones produced by the ovary, one succeeding the other, and each has its special connection with the pituitary.

After menstruation the new egg forms in the ovary and over a period of approximately fourteen days slowly ripens. As this occurs the ovary produces the hormone oestrogen, which reaches the pituitary and the pituitary in turn produces a hormone that stimulates the ripening of the egg. In the middle of the period the egg is shed, ready for fertilization, and then the cavity in which it lay in the ovary is filled with a yellow substance that begins to produce a new hormone, progesterone. But the moment progesterone in the blood reaches the pituitary, it stops producing the first egg-ripening hormone (and the ovary stops producing any great quantity of oestrogen) and now the pituitary produces a hormone which essentially prepares the woman for pregnancy. With the new hormone the mucus lining of the uterus becomes thickened, ready for the embryo to implant if the egg should be fertilized. If no pregnancy occurs, the yellow body in the ovary and the egg deteriorate, and oestrogen begins to be produced once more, informing the pituitary which again assists in the knocking out of the progesterone part of the cycle. This revolution is rather dramatic; the thickened lining of the uterus, so carefully prepared by the progesterone-pituitary relation, is now ruthlessly destroyed and is eliminated, sloughed off, and hence the bleeding of menstruation. If pregnancy has occurred the progesterone phase is maintained throughout the whole gestation period, and some further developments emerge, again involving the pituitary.

The main thing that must be made ready for the birth of the young is the supply of milk. A hormone of the ovary and a hormone of the pituitary work together in harmony to achieve this result; the breasts grow and swell, and after birth the milk flows. These milk hormones, as we saw earlier in this book, have the further property of influencing the mind, the instincts, for it is a well-known fact that in mammals the mother will ferociously guard her infants with a special vigor. Even the calmest and mildest house cat may become a veritable demon after her kittens are born. It is perhaps wrong to ascribe too much importance to the effect of these milk hormones on the mother instinct, since fathers are surprisingly alert in protecting their offspring, and there are many purely visual and auditory cues set off by helpless young in the mind of any adult that normally will produce feelings of sympathy and protection.

Our list of hormones is already long, and we are still far from the end. However, the basic principles involved have been laid down, so we shall just mention a few other important ones, not forgetting that still others have not been mentioned at all. The ovary also produces a hormone during pregnancy called relaxin that loosens the pelvic girdle where it connects with the spine so that the infant can easily pass out of the body of the mother at birth. It is this hormone especially that makes it so awkward for pregnant women to get out of arm chairs, for the rigidity of their pelvis is temporarily lost. Then there are the hormones produced by the placenta which are similar to those of the ovary; the placenta behaves as though it did not entirely trust the ovary and makes duplicate hormones to insure the safety of the embryo. In the male, the small cells between the sperm-producing tubules give off the male sex hormone testosterone, and this again has a reciprocal action with the pituitary. This hormone is not only responsible for the sex urge, that is for mental characteristics, but also during puberty it is responsible in man for the typical hair pattern, the beard and so forth, as well as the deep voice. Of course it is similarly true that in adolescent girls the layer of womanly fat, the breasts, and the hair pattern are in turn the result of the presence of the female sex hormone oestrogen in their blood vessels.

Now let us make a sudden jump and examine the other kind of coordination in animals, nervous coordination. This kind of internal transmission is far more rapid and certainly as complex if not more so than hormone coordination.

The nervous system keeps the body in contact with the outside world. It has, so to speak, sentries posted at key points and these sentries can assimilate certain specific facts about the outside world and pass the information into the internal information headquarters. These sentries are the sense organs, and our senses of touch, taste, smell, and so forth tell us about the immediate surroundings in which we live.

It is interesting that these sense organs are so specific in the kinds of information they can receive. An eye can record only images of light and dark, visual images; an ear can record only sound. As a matter of fact no matter how you stimulate an eye, all it can do is record one thing, and if you are in a pitch black room and have the misfortune to get socked in the eye, you will "see stars." There was really no sudden light, but the brusque banging of the eyeball stimulated the nerves, and there is only one thing they can do, only one thing they can say to the brain, and that is to give an impression of light. Since sense organs are specific in their task, it is easiest to classify them according to the kind of energy to which they respond, be it mechanical energy, or chemical, thermal (heat), or radiant (light).

In human beings the kinds of sense organs or receptors that respond to mechanical energy are numerous. To start with there is the simple matter of touch. Beneath the skin and often associated with the base of small hairs there are delicate nerve endings that respond to the slightest pressure (Fig. 43). The hands and the fingertips, for instance, are loaded with these endings, and as a blind man who can

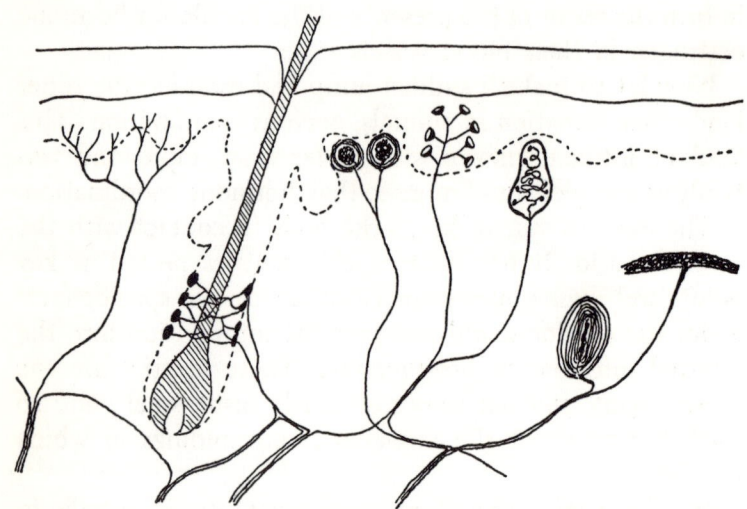

Fig. 43. A section through the skin illustrating a number of the different kinds of receptors found there.

read from a book of Braille knows, they are extremely sensitive and useful as a recorder of the outside world. Lying deeper in the skin there are pressure-tension receptors, and these can be felt as one tightens a muscle or for that matter as one sits in a chair. These receptors are involved when a pilot says that it was a foggy day and he had to "fly by the seat of his pants." By the pushings and pullings of gravity on his body in the cockpit he could tell something of the orientation of the aeroplane.

In the inner ear there are three circular canals oriented at different angles, and these canals contain fluid and are organs of equilibrium (Fig. 44). The effect of gravity on the fluid stimulates sensitive hairs in the canals, and in this way one can keep an upright position. It is easy to disturb

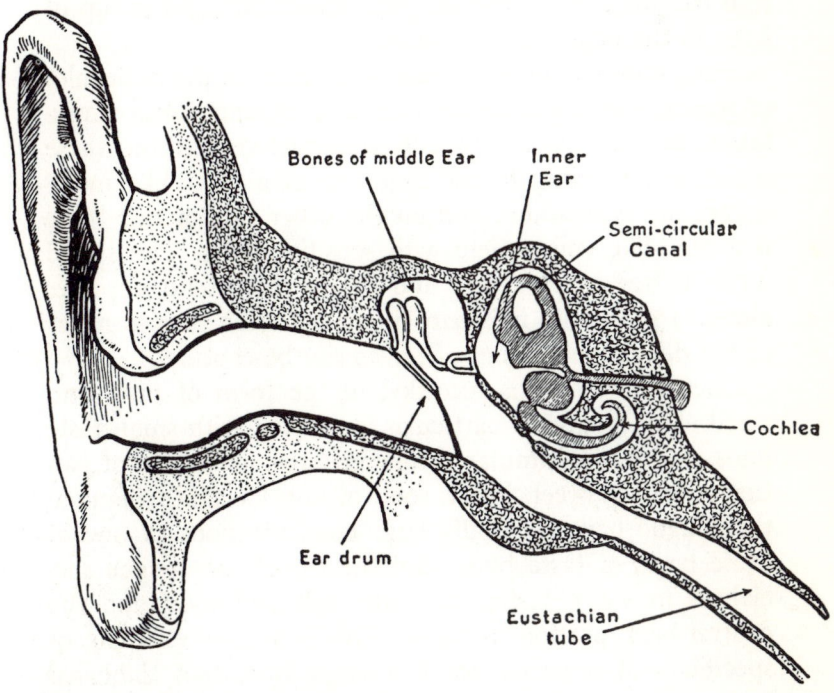

FIG. 44. A section through the ear of a human being. (From H. G. Wells, J. Huxley, and G. P. Wells)

this equilibrium; the child's game of turning about fast many times and then stopping produces dizziness, the feeling that the whole world is turning, and this is simply because the fluid in the ear is still going around even though the body has stopped. The mechanism of these organs of equilibrium can be beautifully demonstrated in certain shrimp that have a simple sac containing a stone that rests upon sensitive hairs. The shrimp will orient so the stone rests on a certain group of hairs, for then it will be upright. At molting time the shrimp take in new stones from the outside, and if bits of iron are supplied instead, the ear stones will be iron after the molt. Then if a magnet is held to the side of the aquarium all the shrimp will swim sideways; they are perfectly happy for they turn so that the iron ball affected by the magnet lies on the right group of hairs in the organ of equilibrium.

Since sound is no more than a vibration of the molecules of the air, hearing is another instance of mechanical stimulation, and the ear is especially constructed to record these vibrations. Leading to the inner ear is a drum-like membrane, the tympanum, and on the other side of this there is a snail-like coil replete with sensitive hairs. The sound waves batter against these and they in turn send impulses along to the brain; it is particularly amazing that so many of the delicate overtones of music can be captured.

Chemical energy is recorded in the form of taste and smell. The tongue in particular is covered with small taste buds, and when stimulated they give the sensations of particular tastes: sweet, bitter, sour, or salt. Certain regions of the tongue have especially large concentrations of one of these types of taste buds, and the tip of the tongue specializes in sweet tastes, the sides salt and sour, and the central back portion in bitter. This is a good example of specificity of response, for if a single taste bud is tickled with a fine camel's hair, depending upon the kind of bud touched, an individual will receive one of the four sensations. In eating, of course, taste alone is not involved, and anyone with a nose-cold is amply aware of this fact for all food appears relatively tasteless; even the finest French dishes taste like baby's pablum. This is because smell is also involved, and the organ of smelling lies in the back of the nose. For a chemical substance to produce a smell, some molecules must fall on the wet surface of the smell receptor and then, by a chemical reaction and a penetration into the cell, the smell can be registered. The curious aspect of smelling is that such small quantities are sufficient to record, and that such delicate nuances between smells can be discriminated by a sensitive nose. I need only remind the reader of Darling's observation of the red deer which could detect human beings down wind for great distances, and again consider how a faithful dog can recognize the smell of his old master even after an absence of many years. Except for famous tea tasters, perfume appraisers, and wine tasters who claim extraordinary feats and can identify the year and the Château of many wines with ease, most human beings lack the extreme sensitivity which many lower mammals possess.

It is easy to tell if water is warm or cold by means of the special thermal receptors in our skin. In fact some are especially sensitive to warmth and others to cold, giving another instance of the specificity involved.

However, of all the forms of energy, perhaps the radiant energy of light is the one which we are particularly well adapted to receive. The human eye is a complex and beautifully designed structure, like a camera, possessing a lens which transmits the image on the back of the eyeball, on the sensitive surface that lies there and is called the retina (Fig. 45). This surface consists of a myriad of special light-receptor nerve cells, and each records its message which is then transmitted to the brain through the large optic nerve chord, and somehow in the brain an image is produced. Not only that, but the picture is in technicolor, for certain of the sensitive cells of the retina have the ability to record colors, while others are involved merely with light and dark, and help when the light is of very low intensity, as in moonlight. For vision, as well as the other senses, I have described what is found in mammals, but it should be mentioned that

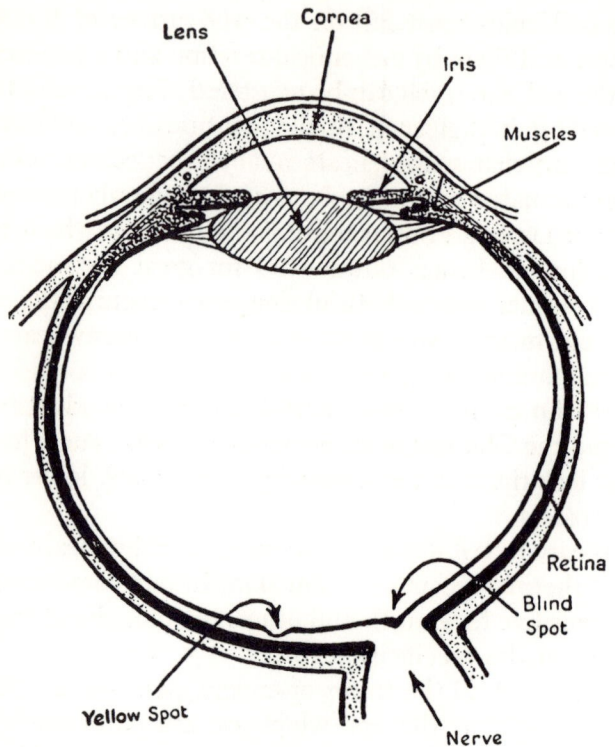

FIG. 45. A section through the eye of a human being. (From H. G. Wells, J. Huxley, and G. P. Wells)

eyes of entirely different construction and origin exist in the different groups of invertebrates, and this is true of the other receptors as well, showing again that the constant feature is the function that is to be performed, and what varies is the means by which this aim is achieved.

Lastly it should be mentioned that there is a special receptor connected with pain. It is of a rather simple construction and found throughout the body, and it is very advantageous to the organism and the species, although admittedly at first glance it would seem to be more of a bother than a benefit. But with the sensation of pain comes a natural instinct of self-preservation, a natural desire to avoid certain dangerous activities that might otherwise be totally destructive. Even in human beings if the sensation of pain is lost in the arm, for instance through some accident, the person must take great care not to burn or mutilate his arm severely without realizing it.

Thus far we have talked only about the part of the nervous system that receives information, and now let us penetrate deeper into the body. In the first place the nerves themselves are interesting structures, for they are cells that have become specialized for the specific purpose of conducting impulses. Like most other cells they have a nucleus and cytoplasm, but the cytoplasm has become thin and elongated in various parts (Fig. 46). Most nerves have a

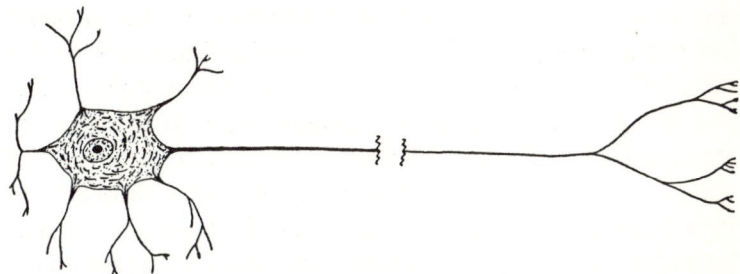

FIG. 46. A typical nerve cell. The nerve fiber is long, and its full length cannot be shown in a drawing of this scale.

rounded portion that contains the nucleus, and on one side of this there are fine feathery cytoplasmic structures that lead out, and on the other end there is one long strand (in some cases in the spinal column it may be several feet long) and at the end of this strand there is again a group of fine feathery extensions. One nerve cell may connect with another in sort of a relay system by having the feathery processes of one interlock with those of another. An impulse starting down a nerve will pass as a wave along its surface, and when it reaches the end of the cell a chemical substance is emitted which bridges the gap to the next nerve, which in turn picks up the impulse and passes it along. At each junction there may be connections with more than one nerve, yet often certain specific pathways are taken, for the internal circuits are well controlled. In fact the chemical strychnine is so poisonous because it destroys this pathway control and all the nerves at all the junctions are stimulated. If a frog, for instance, is given strychnine and then stimulated in any way, it will give one horrible spasmic jump in which every nerve in its body has been short circuited and stimulated, and death immediately follows.

In general there are two main sets of pathways: one coming in from the receptors to the spinal cord and brain, and the other leading out to the effectors, the muscles that do the work. Often the brain is stimulated in this process, but not necessarily so, for certain spinal reflexes such as the knee-jerk reflex occur without any intervention of the brain. But if, while bathing, a crab pinches your toe, obviously sensation plays a part. In any willed action, such as the act of writing, the impulses go from the brain through the nerve chains to the muscles involved. There is then a harmonious interaction between the information-receiving part of the nervous system and the order-giving part of the system.

This coordination is largely a result of the wonderfully intricate properties of the brain and the central nervous system. It serves as a fitting subject for the end of this book because in many ways it is the pinnacle of coordination in any living unit. In social animals the calls and the communication between individuals may be complicated and delicately balanced, but this is absolutely nothing when compared to the mammalian nervous system. Furthermore let us not forget that all the behavior patterns of higher animals in their social activities are possible solely because they are built, as a superstructure, upon the nervous system of each individual.

Recently some further insight has been obtained into the mechanism of the brain, not by biologists but by mathematicians who are interested in electronic calculators, machines that can solve problems, that can think, and that have memory. They boast now of machines that can play good chess and can do many other feats which hitherto were thought to be special properties of human brains, and in fact these machines achieve such personalities that their inventors and operators refer to them fondly by name. Like brains that are made up of a multitude of interconnected nerve cells, they are made up of a multitude of small electronic circuits. But the main factors that utterly defeat the calculator engineer are size and energy, for a large building is usually needed for his machine, and it requires a considerable quantity of electrical power; while animals can do much more with a small brain, and thinking, in a strict physical sense, requires hardly any energy.

It is clear that coordination is not something that living units alone possess, but quite to the contrary, machines and telephone systems have this property often to a marked degree. It is simply one of the essential activities found in organisms, along with many others, and it has been the purpose of this book to show that all these activities are found in all organisms: in societies, in colonies, in cells, or in multicellular organisms. Often the methods by which these activities are performed differ, yet the activity, as well as its survival value in evolution, remains the same. It is easier to characterize life by what protoplasm does than by how it does it.

HUMAN SOCIETIES

In the beginning of this book we started with mammal societies, and then we went down the scale to single cells and from there to multicellular organisms. In every step an increasingly intensive and penetrating analysis was made, and now that we are near the end of the road it may be well to put things back together, and more especially to look back to the place we started, to have another look at a mammal society and see the whole again. The particular mammal society I have in mind is man, for inevitably we compare each social animal and its method of performing the living functions with ourselves and our own activities.

Man is a particular species of mammal and therefore like the howling monkey or the red deer he has need for food and reproduction and coordination. These activities which we observed in the lower mammals were carried out in certain rigid and prescribed ways: for instance, the red deer of Scotland do not differ in their habits from red deer of other regions, nor is there any very great difference between these deer and the elk in our country. Yet man, a single species, shows incredible variation in his pattern of feeding, reproduction, and coordination, over the face of the earth. Every extreme of behavior can be found among far-flung peoples, and this is basically the reason why any comparison between lower animal societies and human societies is studded with traps.

The reason for this profound difference is, of course, obvious. Man has acquired a relatively larger brain, resulting in a more elaborate language and a capacity for relatively more abstract thought. These differences have caused human societies to change quite radically from other animal societies, and we have become to a great extent dependent on customs and mores. By the use of spoken and written language it is possible to accumulate knowledge and to pass from generation to generation all sorts of facts, feelings, superstitions, and fads. This storehouse goes from the father to the son, but not by genes—it is not inherited and passed on to the offspring through the instincts—it is passed on largely by some means of language communication. Human history is so very short when compared to the slow process of the evolution of man which preceded it because evolution through inheritance of variations and the natural selection of these variants takes millions of years, while language-communicated changes and developments can come and go with startling speed.

There is good reason to believe that innate patterns of behavior play a far more important rôle in lower animals than in man. For instance, some weaver birds, which build beautifully elaborate nests, were kept in captivity away from nest-building materials for four generations. Then the fifth-generation birds were again given material, and without faltering they wove beautiful nests. Clearly this elaborate pattern of behavior was inherited. On the other hand it can be shown that many things are learned among animals; a young kitten raised in isolation will not kill a mouse put in a cage with it, but merely play with it gently. But if an old, experienced cat is also put in the cage it will soon teach the young one to kill. In man there are far more actions, like mouse-killing in the cat, that are learned rather than inherited. He is relatively more efficient at learning, and with the help of language there are more things to learn, and from this springs the important consequence that the things to learn can change rapidly with time.

For example, a dramatic change of this sort can be seen in the invasion of western culture on the South Sea Islands or Japan. For many centuries those areas had reached a cultural stability of their own; then within a few years, and not always to their advantage, they learned our ways. In Tahiti their ancient religion was discarded for Christianity, the women were taught the importance (in the eyes of the missionaries) of wearing clothing, and they all learned the pleasure and the evil of gin.

Customs affect not only the diets of people in different parts of the world, but the way in which the food is prepared, the way in which it is served, and the way in which it is eaten. The only common denominator seems to be the biological one involving the necessary amount of food to keep a human body in operation, or more specifically the necessary salts, fats, carbohydrates, proteins, and vitamins to achieve this end.

Likewise, the customs and the laws regulating reproductive behavior vary incredibly in different places, and again the principal common denominator is mainly the biological one of conceiving, bearing, and caring for the children. Concerning this latter point, one striking aspect of man's reproductive activities is that his children need the attention and the protection of the parents for so long a period of time. A child is wholly dependent upon adult care. Part of this is cultural, but surprisingly little. In many countries, including America in rural districts (and everywhere in this country before child labor laws) the children may do or at least help in adult work starting at an early age, but nevertheless they are dependent on the family. The dependency is not entirely economic either, for the children need the love and warmth and authority of adults. This has been shown many times in orphanages and children's hospitals where despite extreme cleanliness and efficiency, children will become sick and a significant per cent die if the nurses do not cuddle them and give them some parental warmth.

The necessity for this long period of care is connected with the fact that the body is slow at growing, but the mind is especially so; and if one compares man with apes and other lower animals, it is found that in general the larger the brain the longer the learning period, and no doubt these two are connected. The human brain is by far the most complex, with many more cells and many more interconnections; no wonder this great structure needs much time to learn and reach maturity.

The human mind is also responsible for our elaborate means of communication and coordination. The mere rapidity of language exchange coupled with the high capacity to learn has been largely responsible for the rapid divergences in the patterns of culture. In western civilization this quick communication has been aided by many mechanical means—the printing press, postal service, the telegram and telephone, radio and television, and these in turn have progressively altered our way of living.

When discussing coordination in earlier chapters we dwelt upon the problem of division of labor, and in human beings this has some especially interesting aspects. It is usual in any human society to have an elaborate division of labor. Not only are the family or household duties divided, but all the activities of the community such as shoe repairing, bread making, clothes making, and clothes selling, watch repairing, caring for the sick, and so forth. All these are done by specific people who concentrate on these activities as professions. But no specific physical type undertakes only one profession. It may be that blacksmiths tend to be stronger than tailors, but if a man is strong he will not automatically become a blacksmith. Even the physical difference between the sexes has not rigidly channeled the general activities of either sex. A man is likely to be stronger than a woman and this usually tends to put him in the place of the protector or the defender of the family. Much of the rôles of the sexes is affected by custom. Witness the rapid rise of women in the last hundred years against their position as chattels, as household drudges whose sole purpose was to cook and scrub for their men and to stand by ever ready to inflate their sagging ego, but the modern wife cannot escape being the one primarily concerned with giving birth to children and caring for them. The respective rôles may vary with the fashions of civilization, but the basic biological differences hold.

The subject of this chapter is endless and fascinating, but my purpose has been only to give perspective to what has gone before and to show how man, in his special way, with the frills of culture added, does basically the same things as other animals and plants. He too is made of protoplasm, and no matter how elaborate is his brain and his social structure, he can never escape from the rules laid down by his body substance.

SELECTED READINGS

SOME GENERAL WORKS ON ANIMAL SOCIETIES
AND ANIMAL BEHAVIOR

(including a few references on bird behavior)

Allee, W. C. 1951. *Cooperation among animals with human implications.* Schuman, N.Y.

Bourlière, F. 1954. *The natural history of mammals.* A. A. Knopf, N.Y.

Collias, N. E. 1951. Problems and principles of animal sociology. In C. P. Stone's *Comparative Psychology*, Prentice-Hall, N.Y. pp. 388-422.

Fox, M. 1952. *The personality of animals.* Penguin Books.

Lack, D. 1943. *The life of the robin.* Reprinted by Penguin Books.

Lorenz, K. 1952. *King Solomon's ring.* Crowell, N.Y.

Tinbergen, N. 1951. *The study of instinct.* Clarendon Press, Oxford.

Tinbergen, N. 1953. *Social behavior in animals; with special reference to vertebrates.* Methuen, London.

Tinbergen, N. 1953. *The herring gull's world.* Collins, London.

MONKEY SOCIETIES

Carpenter, C. R. 1934. A field study of the behavior and social relations of howling monkeys. *Comp. Psychol. Monog.* 10: 1-168.

Carpenter, C. R. 1940. A field study in Siam of the behavior and social relations of the gibbon (Hylobates lar). *Comp. Psychol. Monog.* 16: 1-212.

Collias, N. E., and C. Southwick. 1952. A field study of population density and social organization in howling monkeys. *Proc. Amer. Phil. Soc.* 96: 143-156.

Nissen, H. W. 1951. Social behavior in primates. In C. P. Stone's *Comparative Psychology*, 3rd ed. Prentice-Hall, N.Y. pp. 423-457.

Zuckerman, S. 1932. *The social life of monkeys and apes.* Harcourt, Brace & Co., N.Y.

Yerkes, R. M. 1943. *Chimpanzees.* Yale Univ. Press, New Haven.

FUR SEALS

Bartholomew, G. A., Jr. 1952. Reproductive and social behavior of the northern elephant seal. *Univ. Cal. Publ. Zool.* 47: 369-472.

Bartholomew, G. A., Jr., and P. G. Hoel. 1953. Reproductive behavior of the Alaska fur seal Callorhinus ursinus. *J. Mamm.* 34: 417-436.

Darling, F. F. 1947. The life history of the Atlantic grey seal. In *Natural History in The Highlands and Islands.* Collins, London. pp. 217-231.

Jordan, D. S. 1896-97. *Report of fur-seal investigations.* Treasury Dept. Doc. 207.

Osgood, W. H., F. A. Preble, and G. H. Parker. 1915. The fur seals and other life of the Pribilof Islands, Alaska, in 1914. *Bull. Bureau of Fisheries,* vol. 34.

Scheffer, V. B., and K. W. Kenyon. 1952. The fur seal comes of age. *Nat'l Geographic Mag.* 101: 491-512.

RED DEER AND RELATED ANIMALS

Buechner, H. K. 1950. Life history, ecology, and range use of the pronghorn antelope in Trans-Pecos, Texas. *Amer. Mid. Nat.* 43: 257-354.

Darling, F. F. 1937. *A herd of red deer.* Clarendon Press, Oxford.

Linsdale, J. M., and P. Q. Tomich. 1953. *A herd of mule deer.* Univ. of California Press, Berkeley.

Murie, O. J. 1951. *The elk of North America.* Stackpole Co. and Wildlife Manag. Instit., Washington.

BEAVER

Bradt, G. W. 1938. A study of beaver colonies in Michigan. *J. Mamm.* 19: 139-162.

Morgan, L. H. 1868. *The American beaver and his works.* J. B. Lippincott & Co., Phila.

Seton, E. T. 1909. *Life histories of northern animals.* Scribner's Sons, N.Y.

Warren, E. R. 1927. *The beaver, its ways and its works.* Williams & Wilkins Co., Balt.

INSECTS

Lüscher, M. 1953. The termite and the cell. *Scientific American* 188: 74-76.

Morley, D. W. 1953. *The ant world.* Penguin Books.

Ribbands, R. 1953. *The behavior and social life of honeybees.* Bee Research Assoc. Ltd., London.

Schneirla, T. C., and G. Piel. 1948. The army ant. *Scientific American* 187: 16-23.

von Frisch, K. 1950. *Bees, their vision, chemical senses and language.* Cornell Univ. Press, Ithaca.

Wheeler, W. M. 1913. *Ants, their structure, development and behavior.* Columbia Univ. Press, N.Y.

Wheeler, W. M. 1923. *Social life among the insects.* Harcourt, Brace & Co., N.Y.

GENERAL BIOLOGY

(two books that are both informative and well written)

Thompson, J. A. 1935. *Biology for everyman.* Dutton, N.Y.

Wells, H. G., J. Huxley, and G. P. Wells. 1931. *The Science of life.* Doubleday, Garden City.

ACKNOWLEDGMENTS

I AM extremely grateful and wish to thank the following individuals for their help in reading some or all of the manuscript: H. S. Bailey, Jr., N. E. Collias, J. Hoffman, W. P. Jacobs, and M. J. Levy, Jr.

I should like to thank the following individuals for permission to use their illustrations: W. E. Ankel, G. A. Baitsell, P. Brien, F. A. Brown, Jr., Mrs. W. H. Brown, C. R. Carpenter, C. Drechsler, G. Fankhauser, L. H. Hyman, L. A. Kenoyer, K. W. Kenyon, O. W. Richards, K. D. Roeder, T. C. Schneirla, L. P. Schultz, F. M. Summers, C. L. Wilson.

I should also like to express my appreciation to the following publishing houses, scientific journals, and scientific organizations for permission to use the illustrations listed: American Museum of Natural History, Plates VI, VII, VIII; *Biological Bulletin*, Fig. 15; A. & C. Black, Ltd., Fig. 41; Columbia University Press, Figs. 1, 2; Dryden Press, Inc., Fig. 26; Ginn & Co., Figs. 20, 22, 23, 24, 25; Harper & Bros., Fig. 36; McGraw-Hill Book Co., Inc., Figs. 10, 12, 13, 15; Macmillan Co., Fig. 41; *Mycologia*, Fig. 21; New York Zoological Society, Plate IV; W. W. Norton, Fig. 30; Philosophical Library, Inc., Fig. 4; Smithsonian Institution, Fig. 7; University of Chicago Press, Fig. 11; D. Van Nostrand Co., Inc., Figs. 29, 34; Walt Disney Productions, Plate V; A. P. Watt & Son, Figs. 31, 37, 44; John Wiley & Sons, Inc., Figs. 3, 5, 32; Yale University Press, Fig. 28.

Finally I should like to express my gratitude to Mrs. Marjorie Quick Kolderie for helpful assistance in assembling the illustrations and other clerical help and to Mrs. W. W. Muelken for typing the manuscript.

INDEX

Acrasiales, 104ff.
acrasin, 106
Addison's disease, 209
adrenalin, 209
adrenals, 209
Agalma, 92ff.
Allee, W. C., 20
alligator turtle, 163
ammonia, 130, 191
amoeba, 95, 164ff.
amphibians, 178, 183, 199, 204ff.
amphioxus, 201ff.
Andes, 189
angler fish, 163
animal hormones, 208ff.
antlers, 39ff.
ant eaters, 165
ant lion, 164
ants, 57ff., 67ff., 167, 203
army ants, 67ff.; feeding, 67ff.; reproduction, 70ff.; coordination, 72ff.
ascidians, 204
auxin, 153ff.
Avena-test, 153ff.

baboons, 3ff., 11
bacteria, 96, 111ff., 130ff., 166
Baker, J., 100
Barcroft, J., 189
Barro Colorado, 6ff., 69ff.
bast, 148ff.
bean, 145ff.
Beaumont, W., 169
beaver dams, 51ff.
beavers, 49ff.; feeding, 52, 53; reproduction, 53; coordination, 54
bees, 56ff.
beetles, 167
Bidder, G. P., 175
bile, 170

birds, 165, 179, 183, 192, 199, 223
bladder, 194
blastula, 202
blood, 182ff.
blood pigments, 187ff.
blow flies, 166
bluffing, 10, 28, 41, 54
Buck, F., 103
Boltzman, L., 133
Boysen-Jensen, P., 152

cambium, 146ff.
capillaries, 185
care of young, 3, 12, 18, 29, 34, 43, 57, 224
carnivores, 162ff.
Carpenter, C. R., 7
cell, 95, 115ff.
cell colonies, 95ff.
cellulose, 75ff., 150, 166
cephalopod, 182
cheetah, 162
chemosynthesis, 130ff.
chlamydomonas, 144
chlorophyll, 132ff.
chondromyces, 111ff.
chromosomes, 120, 128, 198
ciliates, 96ff., 115ff.
clams, 165
clothes moth, 192
cockroach, 57, 75
combustion, 125ff., 131
Conklin, E. G., 204
contractile vacuole, 118, 193
cotyledons, 145ff.
cretin, 209
cricket, 57
crustaceans, 175
cybernetics, 221

Darling, F. F., 20, 37ff.
Darwin, C., 20, 62, 143, 167